Report on the Cultivation and Preparation of Tobacco in India
With the Cultivation of Tobacco in Hungary

by Dr. Forbes Watson

with an introduction by Roger Chambers

This work contains material that was originally published in 1871.

This publication was created and published for the public benefit, utilizing public funding and is within the Public Domain.

This edition is reprinted for educational purposes and in accordance with all applicable Federal Laws.

Introduction Copyright 2018 by Roger Chambers

COVER CREDITS

Front Cover
Handmade cigar production, process. Tabacalera de Garcia Factory. Casa de Campo, La Romana, Dominican Republic by Mstyslav Chernov (otrs email by the author)
[CC BY-SA 3.0 - https://creativecommons.org/licenses/by-sa/3.0],
via Wikimedia Commons

Back Cover
Kenbano tobacco leaf by *Agronomist101*
[GFDL - http://www.gnu.org/copyleft/fdl.html]
or
[CC BY-SA 3.0 - https://creativecommons.org/licenses/by-sa/3.0],
via Wikimedia Commons

Research / Resources
Wikimedia Commons
www.Commons.Wikimedia.org

Many thanks to all the incredible photographers, artists,
researchers, biographers, historians, and archivists who share
their great work via the Wikipedia family.

PLEASE NOTE :
As with all reprinted books of this age that are intended to perfectly reproduce the original edition, considerable pains and effort had to be undertaken to correct fading and sometimes outright damage to existing proofs of this title. At times, this task can be quite monumental, requiring an almost total rebuilding of some pages from digital proofs of multiple copies. Despite this, imperfections still sometimes exist in the final proof and may detract slightly from the visual appearance of the text.

DISCLAIMER :
Due to the age of this book, some methods or practices may have been deemed unsafe or unacceptable in the interim years. In utilizing the information herein, you do so at your own risk. We republish antiquarian books without judgment or revisionism, solely for their historical and cultural importance, and for educational purposes.

Self Reliance Books

Get more historic titles on animal and stock breeding, gardening and old fashioned skills by visiting us at:

http://selfreliancebooks.blogspot.com/

Disclaimer

This book was written in an age when little was known about the ill effects of tobacco.

The material presented herein is intended to be strictly for educational purposes with the purpose of enlightening readers about the historical uses of tobacco. Publication of the material is neither an endorsement, nor a criticism of its contents. This book is presented as part of large series of educational material on the history and cultivation of tobacco.

As the reader, please consider it your duty to consult with a medical doctor before utilizing tobacco. It is also the reader's duty to become familiar with local, state, provincial and federal laws relating to the growing of tobacco.

As the author, publisher and retailer cannot control how the reader utilizes the historical information presented in the pages herein, they hereby disclaim any liability to any party for any loss, damage, disruption, death or other liability that may be incurred by the reader's misuse of this material.

introduction

Here at **Self-Reliance Books** we are dedicated to bringing you the best in *dusty-old-book-knowledge* to help you in your quest for self-sufficiency and independence.

We're so pleased to bring you this old title on the cultivation of tobacco. These old publications on agricultural and horticultural topics are ever-popular. It should be said, though, that some of the information is best looked at in the historical context, due to the obsolescence of some practices or methods.

This special edition of **Report on the Cultivation and Preparation of Tobacco in India - With the Cultivation of Tobacco in Hungary** was written by Dr. Forbes Watson, and first published in 1871, making it just shy of one-and-a-half centuries old.

This short, quick read features sections on *Report on the Cultivation and Preparation of Tobacco in India, Tobacco Seed Beds, Planting the Tobacco, Gathering the Tobacco Leaves*, and more.

Another great old book and a must-read for all those interested in the historical aspect of the Tobacco Industry.

~ *Roger Chambers*
State of Jefferson, March 2018

CONTENTS.

	Pages.
REPORT ON THE CULTIVATION AND PREPARATION OF TOBACCO IN INDIA	1–13
MANUAL OF PRACTICAL OPERATIONS CONNECTED WITH THE CULTIVATION AND PREPARATION OF TOBACCO IN HUNGARY	15–63
Choice of the ground and preparation of the soil	15–20
Influence of the character of the soil on the tobacco plant	15
Of materials for manure	17
Tobacco as a rotation crop	19
Preparation of the soil for the cultivation of the tobacco plant	19
Position and protection of the tobacco field	20
Tobacco seed beds	20–27
Laying out of the beds	20
Size of seed beds	22
Quantity of seed for a bed of certain dimensions	23
The sowing of the tobacco seed	23
The tending the beds	24
Transplantation of thinnings or pricking out	26
Planting the tobacco	27–31
Treatment of the tobacco plant till the time of the harvest	31–36
The hoeing	31
Heaping	32
Nipping off the tops and the superfluous leaves	32
Seed plants, harvesting and preservation of the tobacco seed	35
Gathering the tobacco leaves	36–39
Aftergrowth and suckers	39
The drying of the tobacco leaves	40–52
Drying houses for tobacco	40
The wetting of the leaves	43
The stringing of the leaves	44
Close suspensión of the leaves	48
Suspension of the leaves in the sun	48
The drying of the leaves	50
Final suspension	51
The sorting and making into hands of the tobacco leaves	52–58
Table showing the amount and composition of ash existing in the leaves and stems of the tobacco plant	58–59
Description of Plate I.	60
Description of Plate II.	61
Description of Plate III.	62

REPORT

BY

DR. FORBES WATSON,

ON THE

CULTIVATION AND PREPARATION OF TOBACCO IN INDIA.

(1.) The four samples of tobacco raised in India from Cuban seed, and forwarded to this country from Bombay with the letter (No. 34, General Department) of 12th July 1870, have been submitted to two of the principal London brokers in order to obtain their opinion. The tobacco has been condemned by them. Only two of the samples are saleable, and this at the low prices of 3d. and 4d. per lb. The points objected to are: shortness and brittleness of the leaf, attributed to over dryness; uneven colour, and a rottenness of the interior of the packages, attributed to a false mode of packing. These faults, however, are either accidental, as in the case of the wrong mode of packing, or they are preventible by a careful method of curing the tobacco, as will be shown hereafter. *Tobacco raised from Havannah seed in the Bombay Presidency, in the Kaira and Ahmedabad Collectorates, condemned by London brokers.*

(2.) The question of the acclimatization of Cuban and American seed in India, intimately connected as it is with the question how far tobacco can be made one of the staples of Indian export, is far too important to be decided on such insufficient data as a few preliminary experiments can give. The example of the continental countries, France, Germany, and Austria, where the acclimatization of American varieties has been carried out successfully on a large scale, clearly proves that, provided the trials be made in the same systematic and persevering manner, a similar process will succeed in India also, as more favourable conditions of climate frequently prevail there. In the afore-named countries the cultivation of tobacco has made immense progress in recent years, and principally through the action of the governments, which are directly interested in the matter; as in France and Austria the cultivation is a government monopoly. This improvement has been effected partly by the acclimatization of foreign varieties, and partly by an improvement in the methods of cultivation and of preparation generally, in consequence of a closer scientific investigation and the adoption of methods of treating the tobacco more in accordance with its ultimate destination, than the empirical methods hitherto in use. The accomplishment of a similar reform in India will be essentially facilitated by bringing to bear on the Indian experiments this wide range of European experience. *Continental experience in the acclimatization and cultivation of tobacco.*

(3.) As regards Austria there has recently been published by Mr. Mandis, one of the government inspectors for the superintendence of the tobacco cultivation, a valuable guide, describing minutely the method of cultivating and preparing tobacco, which has been found to yield the most satisfactory results. Mr. Mandis speaks from a long personal acquaintance with and experience of the cultivation of tobacco in Holland and in the various provinces of Austria, where tobacco is grown very extensively, about 160,000 acres being devoted to the purpose, and yielding a yearly crop of about 40,000 tons. The conditions and circumstances attending the industry in Austria present some points of resemblance to those occurring in India. The districts devoted to the growth of tobacco in Austria are situated principally in Hungary and Gallicia, where the soil is similar in many respects to that of large districts in India, and the climate very continental and dry. Although the mode of cultivation and preparation is still to a great extent in a very backward state, the quality and quantity of the produce satisfy the internal wants; there, as in India, the improved and extended cultivation is principally insisted on with a view to an export into foreign countries. Here also a large share in the reform to be brought about must devolve on the action of the government, and the principal obstacle is the ignorance and want of agricultural skill and of capital among the tobacco planters. *Manual of tobacco cultivation published by Mr. Mandis, Government Inspector for the superintendence of the tobacco cultivation in Austria.*

The methods recommended by Mr. Mandis applicable to India.

(4.) These considerations lead to the conclusion that a manual destined to meet the demands of the former country will be, to a large extent, applicable to India as well. This conclusion is strengthened by the following fact:—Mr. E. P. Robertson, a very careful and painstaking observer, has arrived, in the carrying out of his experiments on the acclimatization of the Havannah seed, at certain conclusions with regard to the curing of the tobacco in India. Now these conclusions and the principles inferred therefrom are identical with the principles which have throughout guided Mr. Mandis in giving his instructions. Mr. Mandis's manual contains many useful hints towards the practical application of principles already arrived at independently in India. I have, therefore, appended to this report a translation of that portion of Mr. Mandis's manual which refers directly to the practical operations in the course of growing, cutting, and preparing the tobacco. No doubt, a mere mechanical imitation of the directions given will not suffice; what will be wanted is an adaptation of these methods to the modified conditions of Indian Climate.* There will be given hereafter several suggestions as to the nature of these modifications, collected partly from the reports sent by the government officials who undertook to carry out the trials with the Havannah seed, and among which are several reports, in addition to the one by Mr. Robertson, which give proof of great attention having been devoted to the subject.†

(5.) In order to present all these suggestions in a connected manner, and to show the extent and scope of the whole question, it is necessary to condense into a brief compass the leading principles which must be adhered to in order to ensure the success of the proposed experiments. These principles, in so far as inferred from experience in Hungary, will be given on the authority of Mr. Mandis; at the same time, to demonstrate their universal applicability, they will be corroborated by the results arrived at in France, Turkey, and America. It is almost unnecessary to mention that it does not at all enter into the plan proposed to give an abstract of all the scientific knowledge obtained on the subject. The only points touched will be those the direct bearing of which on the practical question of cultivation and preparation has been demonstrated.

The defects of the samples sent from India attributable to bad curing.

(6.) As a preliminary matter, the assertion advanced at the beginning of this Report must be made good, namely, that the faults in the samples presented to the brokers are not due to any radical defect, in the plant, or in the climate of India, but to easily preventible defects of manipulation in the process of curing, and, therefore, that there is ground for the discussion of the subject.

The three defects pointed out, dryness and brittleness, mouldiness, and the uneven colour, are just the defects which Mr. Mandis repeatedly enumerates as consequences of bad drying, as when the drying takes place in too rapid a manner, and whilst the leaf is unequally exposed to the action of the sun and moisture. He especially insists that brittleness and crispness are sure to follow a want of care in the several operations. This circumstance, therefore, cannot be attributed in India to the particularly dry and warm climate, as it occurs very often in the comparatively moderate climate of Hungary. He repeatedly insists that proper care in drying the plant is most essential in order to obtain a saleable article. At page 103 he states, "the plant may have prospered perfectly, and still, by a faulty method of drying, it may become practically worthless."

(7.) Returning, after this digression, to the exposition of the leading principles in the cultivation of tobacco, it will be best to give, in the first place, a view of the results arrived at in Europe on the acclimatization of foreign varieties and the presumable success of similar efforts in India, and then to consider the question of the cultivation of tobacco generally.

The conditions favourable to the growth of Havannah and similar tobaccos are different from those favourable to the production of the varieties usually cultivated in India.

(8.) A few words are necessary to contrast in general terms the characteristics of the tobaccos selected for acclimatization, such as Havannah or Shiraz, &c., from the one in general use throughout India. According to the unanimous statement

---

* Throughout Mr. Mandis's manual there are repeated descriptions of various contrivances for warding off the cold from the plants, such as covering them with mats, boards, shrubs, brushwood, &c. All these passages have been left because similar precautions might be wanted at night during the cold season in parts of India. In a notice published in an Indian periodical, Lieut. W. H. Lowther, writing from Feerozpore, states that in some of his experiments the plants were killed by frost.

Another reason for the retention of the passages in question is, that exactly the same precautions which ward off frost are equally efficient in protecting the plants in their earlier stages from the scorching sun, and are in use in Shiraz, in the Levant, and in India for tobacco and other plants.

† Mr. A. A. Borrodaile, from his letter to the Revenue Commissioner (No. 777), dated 29th April 1870, would seem to have conducted the experiments on a much larger scale than any of the others, although, unfortunately, his detailed account has not been received.

of the reports from India, the chief desire of the native cultivators is to obtain strength, which implies a high per-centage of nicotine in their tobacco. But the Havannah tobacco is exactly that which contains the smallest known amount of nicotine, and the most esteemed varieties of tobacco generally appear to contain less of this principle than the common tobaccos, whilst on the other hand they are distinguished by richness of aroma. It will be pointed out hereafter that the conditions for developing strength are in some measure opposed to those which develop the aroma, and vice versâ. Therefore, in all experiments with new seeds, it is not sufficient to choose what are considered by the ryot good tobacco soils, proper cultivation, &c., it is essential to use other soils and a different culture to develop qualities different from those at present esteemed in the native commodity. Even in India, however, the mode of culture is far from being uniform, although according to testimony the tobacco is usually grown on the richest and heaviest soil. In some districts, as Rungpore,* as also in Java, sandy tracts and rich sandy soils are especially selected for the cultivation; and in Arracan, where one of the best Indian tobaccos is grown, the too rich soils are avoided,† as, probably owing to the excessive formation of albuminous compounds, the tobacco grown on them is not considered good, and will not burn properly. Already long ago‡ it had been stated that in order to obtain in India tobacco suitable for European consumption, the over luxuriance of growth must be guarded against, and that better results have been obtained by selecting poorer soils in a more elevated situation and in a drier atmosphere. These few remarks have been deemed necessary to show that, what are considered favourable conditions for the culture of tobacco by the native farmer in India, are in no way an indication of the conditions necessary for ensuring the successful culture of varieties for the European market, and therefore the new trials ought to be made without too much deference to Indian experience.§

(9.) Exact scientific observations and experiments on the acclimatization of foreign varieties of tobacco are by no means so numerous and conclusive as one would be led to expect from the importance of the subject. They have, however, established beyond doubt, that complete acclimatization,—that is a complete reproduction of the plant with all its distinguishing characteristics,—is not obtained by merely using the seed of the desired variety. The acclimatization depends on a concurrence of several conditions, and in the exact measure in which these conditions are fulfilled the acclimatized variety will approach to or recede from its original character. The observations which bear on the whole question may be considered from a twofold point of view.

1. How far are the botanical characters of the plant, the number, size, and shape of leaves, the position of the ribs, &c., preserved?
2. What are the changes in chemical composition ensuing on acclimatization, as compared with the original plant?

It may be remarked that if the scientific investigation were complete, the two points just mentioned would embrace the whole question. As matters stand, however, the data bearing on these points afford only means for sifting and ascertaining the meaning of the various practical observations made on the smoking qualities of the different kinds of tobacco, on their combustibility, strength, and aroma.

(10.) With regard to the first point, we quote Mr. Mandis (page 59): "Foreign "seed in a given soil will the first year produce a leaf similar to that of the parent "plant, but modified by the local conditions, and if the flowering of tobacco plants "of other sorts in the vicinity be at the same time guarded against, the new seed "will continue to produce the same species of plant with the same form and develop- "ment of leaf as in the first year of its culture; so that between the plant of the

The acclimatised plant preserves its botanical character almost unchanged.

---

* Jour. Agr. Hort. Soc., 25/7, 1846, p. 101.
† Ditto Vol. IX., Part II., 1856, p. 160.
‡ Royle's Essay on the Productive Resources of India, p. 253.
§ High-class aromatic varieties of tobacco are occasionally to be found in India, and the experience afforded by such instances might be turned to account. Cigars of excellent quality are stated to be now manufactured in, or near to, Pondicherry. The following extract refers to the tobacco grown on the island of Cheduba off the coast of Arracan.
"The tobacco is highly praised, and deservedly so. I procured a quantity of it to be made up into "cigars for my own use, and was both surprised and gratified to find among them several of as high "and delicate flavour as any from the Havannah which I had ever tasted, and for the best of which, "but for the manufacture, they might have been mistaken by any one not knowing whence they came." Report on the island of Chedooba by Edw. P. Halsted, Commander of Her Majesty's Sloop "Childers," Calcutta, 1842.

" original seed and its descendants, at least to the fifth generation, after passing
" through this experiment, no difference can be perceived." So far then as the
botanical character is concerned the acclimatized plant assumes from the first year
a slight modification whilst remaining similar to the parent plant, and the new type
thus created does not degenerate further but remains constant.

(11.) With regard to the second point, the observations are very scarce. The
most complete series refer to the composition of the mineral constituents absorbed
out of the soil. What relation these ash constituents have to the organic
compounds of the leaf is not yet satisfactorily established. In fact, as Mr. Mandis
points out (page 36), a full scientific investigation of the necessary connexion
between the various constituents of the tobacco leaf in all its stages of growth,
preparation, and consumption, has yet to be made. The principal facts as
regards the composition of the ash are summed up in the following paragraph.

The absolute amount of mineral substances determined as ash, is very varying in
different varieties, and at different stages of fabrication.

The Table, pp. 58 and 59, contains a résumé of many of the published ash analyses.
In the fresh leaves the amount varies to the extent of 10 to 18 per cent., and more, of
the dry substance; in the leaf, after curing, this amount is larger, and may increase
sometimes to even 27 per cent. of the dry leaf. No less variable than the absolute
amount, is the relative proportion of the ash constituents. Whilst in No. XIII. in
the Table, the potash constitutes more than one half of the basic elements present, in
other instances the amount of lime and magnesia is so great that potash forms
hardly one sixth of the bases. Besides the mineral compounds found in the ash,
there are others destroyed or volatilized in the process of incineration; these are,—
nitric acid, probably combined with potash to form nitrate of potash, and chloride or
nitrate of ammonium, together with the chlorides of potassium and sodium. The
analyses referring to these two important substances are rather contradictory.
The amount of nitric acid has been found by Mr. Schlossing to vary between 0·15
and 6·1 per cent. in the midrib, and between 0·2 and 2·1 per cent. in the stripped leaf.

The chemical composition, as far as dependent on the mineral constituents, changes with the nature of the soil.

(12.) The influence of acclimatization on the composition of the ash is apparent
from the following table, which embodies the results of an experiment carried out
by Dr. Kodweiss. It consists in the comparison of the ashes of the leaves of two
different varieties, a Hungarian variety from Debrö and an original Virginian variety.
The seeds of these two varieties were sown at Heinburg in Hungary in the same
soil, and the ash of their leaves analyzed.

Constituents.	Original Debrö. Prepared Leaf.	Original Virginia. Prepared Leaf.	Debrö, grown at Heinburg. Green Leaf.	Virginia, grown at Heinburg. Green Leaf.
	Per Cent. of the dry Leaf.			
Potassa	2·694	4·666	1·730	1·633
Lime	5·420	2·296	7·554	7·078
Magnesia	2·330	1·114	0·670	1·832
Oxide of Iron	0·370	0·300	0·418	0·188
Silica	0·040	0·050	0·090	0·110
Chlorine	1·490	0·466	0·216	0·626
Sulphuric Acid	0·414	0·517	0·980	0·820
Phosphoric Acid	0·749	0·656	0·732	0·699
Total Ash	13·507	10·065	12·390	12·986

This table is of much interest. The first two columns show the considerable
difference which existed in the composition of the original leaves, the Virginia leaf
containing almost double the potash of the Debrö leaf, and only half the lime. By
columns three and four it is seen that when the seed of these two varieties was
sown upon the same soil the composition of the ash was strikingly similar. The
differences observed may be due to the varying composition of different leaves.
The inference from this experiment is, that tobacco plants grown from different
seeds on the same soil will possess substantially an identical chemical composition
as regards their mineral constituents.

(13.) There are no similar comparative experiments in existence relating to the
organic compounds of the leaf generally, nor even to the active principles of
tobacco. What data are known on this subject are either unconnected and often

merely approximate determinations of single constituents, or inferences drawn practically from the different behaviour and properties of the prepared tobacco during smoking. These data will therefore be but mentioned in the following paragraphs together with the observations referring to the third point, namely, to the practical test of the tobacco, by comparing the combustibility, strength, and aroma of the prepared leaves and the influence of acclimatization thereon.

(14.) All that the present state of our knowledge allows us to assume is this. The individual and distinctive characters of the various tobaccos, as articles of consumption, depend mainly on the proportion of four elements:— *[Conditions on which the value of tobacco, as an article of consumption, depends.]*

    1st. On those of the mineral constituents, among which the potassium salts are most important:

    2nd. On the amount of albuminous compounds:

    3rd. On the amount of nicotine:

    4th. On the amount of nicotianine and of the essential oil.

(15.) The influence of the mineral constituents on the properties of the tobacco leaf, as at present established, seem to be twofold. In the first place, the ash seems to act by its great amount as a preserving and antiseptic principle, preventing and stopping the fermentation, thus facilitating the operation during the curing of the tobacco, and rendering the final commodity stable and unalterable.* In the second place, the presence of a large quantity of potassium salts, and especially of the nitrate and carbonate, seems essential to assure a proper burning of the prepared leaf,—a most important point in the estimation of various kinds of tobacco.† *[Potassium salts favourable to the combustibility of tobacco.]*

In so far then as these two properties are concerned, the value of the products will not depend on the kind of seed taken, but only on the soil on which the plant has grown and the kind of manure which has been employed, as the experiment quoted above proves that plants from different seeds raised on the same soil will contain a similarly constituted ash. Whether the amount and quantity of ash have a distinct influence on the active principles of the tobacco is as yet only a matter of conjecture; it is however probable that they have.

(16.) The albuminous substances amount to 10 and more per cent. of the dry leaf before curing. Their presence in the prepared tobacco prevents the proper burning, and besides gives rise in burning to a disagreeable smell which overpowers the aroma of the leaf. It is the chief end of curing and fermentation to get rid of these substances. The relation which the different varieties of the plant, or the different nature of the soil, bear to the amount of albuminous matter, is not yet ascertained. *[Albuminous substances unfavourable to the combustibility of tobacco.]*

(17.) The nicotine is the active principle of the tobacco which constitutes its strength. Its amount has been variously estimated in different varieties. Schlössing found in a French tobacco 3·8 per cent., once 9 per cent., in Kentucky 6 per cent., in Virginia 6·8 per cent., and in Havannah not quite 2 per cent. The better varieties generally contain less of this principle than the usual varieties. A fertile heavy soil, strong manure, and abundance of moisture are the conditions which facilitate its formation. Besides, the amount varies very much with the period of gathering the leaves; it increases as the leaves become ripe. *[Strength and aroma of tobacco, and the conditions favouring the one or the other.]*

The conditions which favour the formation of the compounds which give to the tobacco its aroma are just the reverse,—they are sunshine, warmth, and a light airy soil, sandy or calcareous. Thus it comes to pass that a given variety of tobacco may either be very strong, but containing fewer of the aromatic principles, or rich in aroma and less strong, the latter kind being the more esteemed of the two. The Havanna is the example *par excellence* of a tobacco of this kind. According to Dr. Kodweiss the aromatic substance consists of two different bodies; of nicotianine, which is found uniformly in all varieties of tobacco, and of an essential oil characteristic of every special variety. Thus while it is doubtful whether the amount of nicotine depends at all on the variety of *the plant*, it seems that in every variety the aroma is in some degree a distinctive property.

(18.) From this cursory view of the principal constituents of the prepared tobacco leaf, it appears that its outward shape and size, the strength and disposition of the ribs, &c., together with the character of the aroma, depend to a large extent on the kind of seed used, and will probably remain substantially unchanged in the acclimatized plant; whereas the amount and relative proportion of the mineral constituents depend exclusively on the soil and its cultivation; and the amount of

---

    * Mr. Mandis, p. 92.
    † M. Barral. Rapport au Jury de l'Expos. Univ., 1867.

nicotine and the quantity of the aromatic substances partly on the soil, and partly on the joint influences of temperature, sunlight, and moisture.

<small>Decisive influence of the soil on the character of the acclimatized tobacco.</small>

The conclusion to which the above paragraph more especially points is the importance of the influence exercised by the soil. It is obvious that in all experiments on acclimatization, the comparison of the acclimatized plant with the original variety is quite inconclusive, as long as the nature of the soil is not taken into account, as even in the same country and under the same climatic conditions it is impossible to raise plants similar to the parent plant, on a soil of a different description from that on which the parent plant grew. Numerous experiments and observations bear out this view. Mr. Mandis gives (page 40) an account of an experiment tried on a large scale in Hungary. There are in that country, among others, two very different varieties of tobacco, the tobacco of Debrö, grown on a sandy soil, and distinguished by its aroma, and the tobacco of Szegedin, grown on a heavy fertile soil, and distinguished by its strength. If, however, the Debrö seed be sown in the neighbourhood of Szegedin, the plants will be found to lose every trace of their Debrö origin, and to differ in no respect from the native Szegedin plant. Among other examples, the Report of the Turkish Commissioners for the Paris Exhibition of 1867 points out that the characteristic properties of the tobacco grown at Salonichi (Yenidje Kareson) are seemingly dependent on the soil, as in neighbouring districts, the mode of cultivation and preparation being alike, the tobacco is widely different. In the numerous notices about Indian tobacco published in the proceedings of the Agri-horticultural Society of India there is repeated mention of a valuable variety of tobacco restricted in its growth within narrow limits, and different from the usual kinds grown in the villages around.

(19.) The action of the soil on the plant is twofold. It acts by the chemical constituents contained in it, and by its state of aggregation and its physical properties. As the tobacco is a very exhaustive plant, it wants an ample and rapid supply of its ash constituents and of ammonia. The want will be best supplied when the soil contains a great proportion of vegetable mould, as this will present a large proportion of the ash constituents in a soluble form. The physical properties of the soil which most influence the cultivation of tobacco are, its state of cohesion, its power of retaining water, and its power of absorbing heat. For the cultivation of aromatic varieties of tobacco, a light loose soil, readily absorbing heat, is required, such as a sandy or calcareous soil. This kind of soil will never have a high retentive power for moisture, and this is of considerable importance, as stagnant moisture must be carefully avoided. On the other hand, the soil should always remain slightly humid. A combination which, according to Mr. Mandis, satisfies these seemingly contradictory conditions is, for instance, a light sandy or calcareous soil, with a clay subsoil. (This is only noticed here as an instance of the numerous conditions which must concur to ensure the attainment of first-rate qualities of tobacco.)

<small>Influence of climate,—that is, of sunshine, heat, and moisture,—on the character of the tobacco.</small>

(20.) The next important conclusion is, that in order to obtain the same combination of strength and aroma in the acclimatized plant, it must be placed under equally favourable conditions of temperature and moisture. In this respect India is situated much more favourably than the European States, where the foreign varieties of tobacco have been acclimatized, and where the principal advantage consists in the better shape of the plant, and in the position, number, and disposition of the leaves. In Europe the climate is such as not to allow of the full attainment of the original aroma; although, even as regards aroma, the plant raised from Cuban seed in Austria is favourably distinguished (Mr. Mandis, page 56). The diversity of climate in India ought to enable us to put the acclimatized plant under conditions similar to those existing in the country where it is indigenous.

(21.) The Havannah tobacco, and other aromatic varieties in general, require in their latter stage of growth the full effect of sunlight, in order to develop the aromatic substance; at the same time a plant so leafy and bulky as tobacco will always want a considerable amount of moisture. The combination of sunlight, warmth, and moisture, most suitable for bringing out the fine qualities of the Havannah seed, may be found in two ways; firstly, by the choice of a district where the climate is in all respects most similar to the Cuban climate. Mr. Robertson suggests in his report that cultivation be tried in Guzerat, which appears to him to come nearer to the Cuban climate than the other portions of India. Another mode of attaining the object in view, and one not restricted in its application by geographical limits, is to vary the climate of a locality with respect to the tobacco, simply by varying the season in which the plant is grown. Of the climatic elements enumerated above, the sunlight and temperature are beyond human interference, and therefore the period of cultivation must be chosen in such a manner as to afford plenty of these two elements during the whole period that the tobacco is ripening; when the general climate falls short with respect to moisture the deficiency may be rectified by artificial irrigation or watering. Mr. Robertson suggests a change in the season of cultivation for another reason, but his idea may be amplified by instituting comparative experiments, by planting tobacco perhaps every month or every two months, each time performing all the successive operations in their proper cycle, time, and order, and at the end of the year comparing the tobaccos grown in these different periods.

(22.) The previous paragraphs have treated of the general question of the acclimatization of foreign varieties of tobacco in India, but supposing this acclimatization carried out successfully, there remains another problem to be solved. A system of cultivation and of preparation must be introduced which will bring out to full advantage the intrinsic qualities of the plant, and allow of the production, in a commercially advantageous manner, of a technically faultless commodity, satisfying the exigencies of the European markets. This system will apply not only to the acclimatized varieties, but also to the tobacco plant generally. *[margin: Conditions necessary for ensuring the success of tobacco cultivation generally.]*

The cultivation and preparation of the acclimatized as well as of the native varieties of tobacco can only be carried out successfully when certain fundamental principles are acted on. The most important of these refer to the following points:—

1. Choice of seed.
2. Proper system of manure and of rotation of crops.
3. Proper system of cutting and gathering the ripe leaves.
4. Proper mode of curing the leaves.
5. Commercial assortment of the produce.

(23.) 1st. As regards the seed, it may be observed that in France* the greatest attention is now paid to its selection. Formerly, nearly every cultivator of tobacco provided his own seed; now, the government administration has taken this matter into its hands, and grows its own seed, selecting with extreme care only the finest plants for it, and this seed it supplies to the farmers, who are prohibited from using any other. There are several advantages in this system. The seed proceeds exclusively from the very best varieties, and each variety is kept distinct in cultivation, so that the seed sown by the farmer is uniform in kind. This is a very important matter. Formerly the seed employed was almost always a mixture, so that the plants were very often of undesirable crossings, and did not ripen at the same time; and, besides, in the further processes of curing and fermentation, a great deal of trouble was entailed in sorting the different leaves, which, being of different varieties, wanted different times for the same operations. All these circumstances, rendering it difficult to obtain finally an uniform and homogeneous article, have now been obviated in France. The case, however, is different in Hungary, where Mr. Mandis considers that the carelessness with respect to the seed employed constitutes a chief fault in the system of tobacco cultivation as practised in that country. *[margin: Choice of the seed.]*

(24.) 2nd. A proper system of manure and of rotation of crops is of great importance, because tobacco is a very exhaustive plant. The analyses in the table at p. 58, show that the mineral substances essential for the growth of tobacco are chiefly bases, —potassa, lime, and magnesia,—whereas the amounts of phosphoric, sulphuric, and silicic acids are less important. Now as these latter substances are exactly those most important in the cultivation of grains the position of tobacco in the rotation of crops is fixed. Of the store of available mineral substances in the soil, made up partly by the ever progressing decomposition of its constituents and partly by manure, the tobacco will principally exhaust the potash, lime, and magnesia only, whilst phosphoric, sulphuric, and silicic acids will go on increasing in amount. If now a crop of grain be taken from the same soil these acids will be used up, whilst potash and lime will be accumulated, thus restoring to the soil the conditions for growing tobacco. Mr. Mandis (page 49) gives some directions relative to the practical application of the above results, and, besides, points out that the continuous growth of tobacco on the same field has some collateral disadvantages in addition to the exhaustion of the store of bases. A proper manure is of essential importance—and of course a manure which will restore to the soil the substances taken up by the tobacco. Thus, *[margin: Rotation of crops and careful manuring indispensable.]*

---

* M. Barral, in Jury Reports for the Paris Exhibition of 1867.

on soils poor in lime the use of burnt lime or gypsum is recommended. Cow, sheep's, and goat's dung are most usually employed, besides sewage, which, being rich in ammonia and potash, does very good service.

The gathering and cutting of the tobacco ought to be done at several successive periods, selecting each time only the perfectly ripe leaves.

(25.) 3rd. Next in importance is a proper system of cutting and gathering in the ripe plant.

Carelessness in this respect is one of the greatest defects in the cultivation of tobacco as practised in India. Throughout Europe, in the United States, in Cuba, and in the districts where the best Turkish tobacco is grown, the gathering does not take place at once, but extends over a long period, the leaves being taken one by one as each gets ripe. The importance of this care will be apparent from the following remarks. The leaves of the tobacco do not all ripen at the same time. There is considerable diversity in this respect between the different varieties. In Holland, Germany, and Hungary, three principal classes of leaves are distinguished; the lowest and smallest leaves, ripening before all others, are distinguished by the Dutch name "Sandgut." The next are the middle leaves of the largest size, called "Erdgut," ripening a week or two after the lowest. The highest leaves are again smaller, and are called "Bestgut," and ripen about a week after the "Erdgut." Thus, the gathering mostly extends over from two to three weeks, and is repeated three times successively. Some varieties of tobacco require only a double gathering, whilst others are taken in at four, five, and six different periods, the leaves on the transition from sandgut to the erdgut, called "Lumpsel," and the leaves between the erdgut and bestgut, called "Zweifler," and finally the top leaves, called "Spizblatter," being all collected separately. The same care is taken in Turkey to collect only the ripe leaves. In Albania and in the district of Salonichi, where the finest Turkish tobacco is grown, the "yenidje karason" variety, the gathering extends over from three to four weeks, and takes place at five different periods, beginning with the lowest leaves ("dib yaprak"), then the next above ("dib ustu" or "dib kabassi"), then the middle leaves ("orta direm" or "bayuk ana"), then the upper ("ondj alti" or "ikindji ana"), and finally the topmost leaves ("ondj" or "kutchuk ana"). If the gathering of all the leaves is done at the same time, then one cannot fail to gather one portion of them while yet unripe, and another portion when over-ripe. The consequences are almost equally fatal. This is a translation of the passage in which Mr. Mandis refers (page 92) to these consequences: "Leaves taken whilst still unripe, become uneven in " colour, thin and powerless after the drying, and as they are not yet fully " saturated with mineral substance, they exhibit through repeated and violent " fermentation a great disposition to decompose and yield a tobacco of inferior " quality, nay, often quite worthless. The over-ripe leaves also lose in value, " because the absorbed substances are again carried downwards with the descending " vital sap, and diminish in consequence of the exosmose; the substance of the leaves " becomes porous, loses in elasticity, shows a greater tendency to mouldiness, and at " the least assumes an unpleasant light colour."

(26.) The theoretical reason of this behaviour of the leaf consists in its being an ever varying substance, the processes of combination and decomposition, of endosmose and exosmose going on uninterruptedly, with the prevalence of the one or the other at different times. The proper moments for the gathering of the leaves are just the latter stages of ripening, when the mineral substances are rapidly increasing. Then the mineral substances begin to diminish again, thus reducing the combustibility of the final product; the gummose substances diminish equally, which renders the dried leaves less elastic and more crisp and brittle, and subject to being reduced into dust; and finally the proportion of nicotine is rapidly increasing, although in all finer qualities of tobacco an excess of it ought to be carefully avoided. It is for all these reasons that a careful and early gathering is enumerated by Mr. Barral in the Jury Report of the Paris Exhibition in 1867, as one of the principal reforms introduced into the French system of tobacco cultivation.

Successive operations included under the general term "curing."

(27.) 4th. The curing of the leaves is perhaps the most important operation. It can be done properly, only when the previous operations have been executed with all the care insisted on in this report, for even the best material can be entirely spoilt by bad curing. Curing consists of a series of operations. The cut leaves are first allowed to wilt, in the next stage they acquire the proper colour, then they are dried, made into "hands," and finally undergo a fermentation. In the first three stages most of the usual methods of manipulation are defective. It is here to be noticed that in the course of his experiments in Dharwar, Mr. E. P. Robertson hit on the right principles, and Mr. Mandis' detailed directions will show the full practical application of them, although no doubt some of his prescriptions ought not to be followed servilely, but ought rather to be adapted to Indian conditions. Notably, the times required for the various operations may require considerable variation.

(28.) The leaf is subjected to a remarkable transformation during the curing. The organic substance undergoes the process of decomposition; water and carbonic acid are given off, and compounds are formed containing a higher per-centage of carbon, distinguished by a brown colour, and probably analogous to the brown compounds composing the mould produced by decay of vegetable matter. At the same time the albuminous substances are being partially destroyed during the whole process of curing, and especially during the final stage of fermentation. *Chemical changes during the process of curing.*

(29.) This chemical transformation is a gradual process, and requires time, weeks and months even, to develop itself fully, and during this time there are other influences at work which may become injurious to the product if the utmost care be not taken. 1st. If the leaves are allowed to dry too soon, and especially if they are exposed to the sun, the process of the decay of the substance of the leaf, and the decomposition of albuminous matter is left incomplete; some of the shaded portions of the leaves remain green, and the portions exposed to the sun get yellow, not in consequence of the internal decomposition, but solely because the chlorophyl, or green colouring matter, becomes bleached by the sun. Such a leaf will finally present an uneven colour, a chequered appearance, and, especially the portions which got dry in the sun, will be very brittle and crisp. 2nd. If, in consequence of careless manipulation in the handling, the leaves are allowed to rub one against the other, or if moisture in drops collects on their surface, either by rain or by artificial moistening, or even by too violent a sweating, then those places begin to rot, become very deep brown or even black, the fibres of the leaves become injured, and the leaves altogether become brittle after drying. 3rd. The fleshy midribs are a great difficulty in complete drying, and unless they get completely dry they will entail mouldiness, which may communicate itself to a great portion of the leaf. 4th. The operations during the curing require a repeated handling of the leaves, and only a very methodic way of manipulation will prevent mechanical lesions, holes, and fissures, all which cause a serious diminution in the value of the produce, because such leaves cannot be applied to the manufacture of cigars. *Enumeration of the defects due to bad curing.*

(30.) It will be seen that most of these points are taken into account in Mr. Robertson's directions, which are,— *Conditions which must be fulfilled in order to ensure good curing.*

1. If the tobacco is to be cured properly there must be no haste,* and no attempt whatever at artificial wetting of the leaves.
2. The colour must be obtained before allowing the leaves to dry.
3. Tobacco should never be made into hands to undergo fermentation until the centre stem of each leaf is perfectly dried up.
4. The dry tobacco is to be carefully kept until the next rains; then only will it bear handling, and be able to be made into hands previous to the fermentation.

(31.) 5th. The proper manner of sorting, packing, and arrangement for the market generally is next to be considered. These operations, which are performed at various periods during the curing of the tobacco, and usually terminate definitively at the end of the drying, cannot, of course, influence the essential quality of the produce. They are, however, of very great importance in ensuring commercial success. The sorting must have reference to three different objects; 1st, it must refer to the ultimate destination of the tobacco, so that tobacco suitable for the production of cigars shall be separated from that which is to be used for cutting-up and from that used for the manufacture of snuff; 2nd, it must look to quality, that is to the more or less successful curing, so that all the kinds of leaves are again subdivided *Assortment of the produce for the market.*

---

* The brittleness of the cured leaves has been ascribed to the effect of the great dryness of the Indian climate. The explanations given in this report show that brittleness has many other causes besides over-dryness. Nevertheless it is possible that evaporation will go on so rapidly in India that it will not allow sufficient time for the chemical transformations during the curing. Mr. Robertson suggests, although with a different end in view, that an artificial damp atmosphere might be obtained in the tobacco rooms or houses by the use of wet kuskus or jawasee tatties. Straw or other similar materials spread on the ground and sprinkled over with water might produce the same effect.

Mr. Robertson in one of his observations found that the tobacco after drying was very much improved by being kept until the next rains before being made into hands. The operations described in Mr. Mandis's manual under the head of "final suspension" allow the leaves to be kept in an unchanged condition for a considerable period, until the occurrence of rain or other states of the weather suitable for farther manipulation.

into three or four portions of different quality, by which means a much better price is obtained for the whole quantity of tobacco, because, if a small per-centage of good leaves be interspersed among a large number of second-rate quality, the manufacturer will disregard the good leaves altogether, and fix the price as if the whole bale were uniformly second-rate; 3rd, it must take size into consideration.

Good sorting is a most tedious and difficult operation, and can only be carried out successfully when the precautions here insisted on have been observed during the whole cultivation of the plant, that is, when the same kind of seed has been employed and the plants consequently all belong to the same variety, and when the gathering of the leaves has taken place in such a manner that the three to six different kinds found on the same plant have all been collected at different periods and kept separately.

Besides, it needs a very good judgment to recognize the precise quality of every leaf and its special suitability for some particular destination, a destination often dependent not only on the inherent qualities of the leaf but also on arbitrary trade customs. In the appended translation of Mr. Mandis' manual there is a detailed description of different varieties of leaves and of the special elements which influence their value. The best description, however, and the most careful plates will fail to impart more than some general information on a subject, in which a scrupulous exactitude in details and adherence to minutiæ is wanted.

This assortment can only be made satisfactorily by providing trade-samples, accessible to the producers in India.

(32). It will hardly be contested that the only satisfactory information will be an ocular one obtained by providing and sending out to India trade-samples of the different varieties and qualities of tobacco. Only by having such collections of trade-samples of this and other produce accessible to producers in India will they be able to satisfy the exigencies of the European market; whereas, otherwise, a defect in mere form might prevent the value of the article from being rightly appreciated. The question of trade-samples touched on here is, however, only a portion of a scheme to which I shall refer in another communication.

The end to be kept steadily in view is the production of prime qualities.

(33.) From the exposition presented in this Report, it is manifest that great and systematic care must be given to the acclimatization and preparation of tobacco to ensure a good result. From the choice of the district and soil where the plant is to be grown, through all the stages of cultivation and preparation, every stage is of decisive influence on the final produce, and the neglect of any one of the manifold precautions will at once tell upon the marketable value of the leaf, and render the production unremunerative. It is obvious that repeated trials will be necessary in order to produce a first-rate merchandise, and it is only to the production of first-rate qualities that one must look in the fabrication of tobacco for export. This is the same advice which Mr. Mandis gives to the Austrian cultivators. He insists that the cost of packing and of transport is so great, and the competition in the inferior qualities so strong, that their export will seldom pay. This refers to the tobacco destined to be cut or prepared into snuff, but it applies in a still higher degree to tobacco suitable for the manufacture of cigars; these leaves fetch comparatively the best price, and they must be the most faultless. As an example of how much depends on care and method, he states, that whereas in Hungary the leaves suitable for the manufacture of cigars form only 5 per cent. and sometimes 10 to 15 per cent. of the total weight of produce, the Dutch usually sell as much as 50 per cent. as cigar leaves. The production of prime quality must be, therefore, the end kept in view, and in this manner it will be possible not only to produce acclimatized varieties of tobacco fit for European markets, but, by force of example, the whole native cultivation will be brought to a higher degree of excellence.

The introduction of a system of cultivation and preparation of tobacco, yielding first-rate qualities, involves so much care that it can only be successfully attempted by means of experimental farms.

(34.) The very wide scope of the question is evident. It involves nothing less than the reform of a considerable branch of agriculture, a reform not only profitable from an agricultural point of view, by raising the value of one of the Indian staples, but also salutary from hygienic considerations, by substituting the more aromatic and enjoyable varieties of tobaccos for those of which the strength lies in their nicotine, a substance which cannot fail to exercise a deleterious action on the system.

This reform can only be brought about gradually and by the force of example, which is the only manner in which agricultural reforms have ever been introduced. The possibility of growing superior varieties of tobacco must be demonstrated practically, and this not only in a few garden experiments, but on the same scale and in the same manner as it is to be conducted by the producers. I fully concur in the observations of Mr. G. W. Elliot, in the 5th paragraph of his letter to the Government of Bombay (No. 215 of 1870), to the effect that this can only be accomplished by means of farms under the charge of practical agriculturists who can give to the subject the full care it requires.*

(35.) I need only allude to the success with which the same means have worked in the case of the cultivation of tea, and are now being applied to the cultivation of cotton and other commercial plants.

Tobacco belongs to exactly the same category, and the general scheme followed in the efforts made in introducing and improving the cultivation of other staples applies equally to it.

Experimental farms have hitherto proved successful in attempts of a similar kind.

The three leading features which schemes for such objects have always presented are :—

1. Improvements and selection of the species, including seed farms for placing within the reach of the agriculturist the seed of the most favourable species.
2. Improvements in the cultivation of the soil and the tending of plants, in order to obtain a maximum yield per acre.
3. Improvements in the different processes by which the raw natural produce is converted into a saleable commodity, such as ginning and pressing in the case of cotton—drying, roasting, and rolling, in that of tea—and drying and curing in that of tobacco.

(36.) The proposed measures are, therefore, not merely tentative, and promising only a problematical success; on the contrary, precedents applying very closely to the point show that it is only necessary to follow a known track and apply principles which have already succeeded in the case of other commodities.

Similarity between the cultivation of tobacco and that of tea.

And the similarity between the proposed cultivation of tobacco and that which has been successfully carried out in the case of tea is very great. In both instances the object aimed at is the production of a leaf containing certain active principles, and combining certain conditions as regards strength and aroma; and even the processes in the final preparation of the leaves are in some measure analogous, and certainly require as much nicety of manipulation in the one case as in the other.

(37.) More than this, the conditions under which the experiments on the tobacco will have to be conducted are far more favourable to their success and to their economical importance than was the case in the parallel instance of tea.

Tea was an entirely new culture; the popular interest in it had to be created; whereas, thousands of acres are already devoted to the cultivation of tobacco, and a vast mass of people are already interested in everything which affects the commerce in one of their own chief articles of production. It remains only to open to them the prospect of an advantageous export trade by showing the preliminary conditions which must be satisfied in view of this prospect.

There is little doubt that if once the more enterprising agriculturists adopt improvements in order to produce an article of export, their example will spread, and that the whole cultivation of this article will, in time, be put upon a new footing.

(38.) In the foregoing paragraphs the subject in question has been regarded from a general point of view.

The bearing of the proposed measures and the principles which must be observed in their execution have been discussed, and reference has been made to experiments —in every respect similar to the methods of cultivation here recommended in the case of tobacco—which afford examples of the practical organization necessary to render success probable.

Independently of this, however, I venture to draw attention to the proposal of Mr. Stuart Clark, Inspector General of Prisons in the North Western Provinces, to attach farms to each of the central gaols.

Proposal to establish farms in connexion with central gaols.

The connexion of this proposal with the whole system of industrial and agricultural labour in gaols will form the subject of another communication. I only advert to this matter here because the establishment of such gaol farms would afford especial facilities for the organization of experiments for the successful cultivation of tobacco and other products.

(Signed)   J. FORBES WATSON,
Reporter on the Products of India.

---

* Since writing the above, my attention has been directed to the proceedings of the Madras Government for the month of October last (1870), which show that a step in the right direction has been taken by the Government of that Presidency, which has sanctioned the suggestion of the Board of Revenue, to the effect that the experimental cultivation of tobacco should be at once commenced, under Mr. Robertson, Superintendent of the Government Farm, and that the results should be submitted to Mr. Broughton for analysis.

A MANUAL

OF

PRACTICAL OPERATIONS CONNECTED WITH THE CULTIVATION AND PREPARATION OF TOBACCO IN HUNGARY.

Extracted from the "Anleitung Zur rationellen Tabak Kultur"* of J. Mandis, Financial Counsellor and Inspector for the purchase of Tobacco in Austria.

CHOICE OF THE GROUND AND PREPARATION OF THE SOIL.

INFLUENCE OF THE CHARACTER OF THE SOIL ON THE TOBACCO PLANT.

The character of the soil, that is the mixture of its ingredients, exercises an important influence on the growth of the tobacco plant and on the quality of its produce.

The tobacco plant draws from the soil, for the formation of its substance, carbonic acid, ammonia, lime, potash, magnesia, oxyde of iron, chlorine, silica, sulphuric acid and phosphoric acid; these substances must, therefore, be present in certain quantities, and in a soluable state, for if but one of them be wanting the plant will not attain perfection, and all the other nutritive matters will remain without effect.

These essential ingredients are indeed present in most kinds of soils, but their quantity and quality must be in suitable proportion to the wants of the plant, otherwise it cannot properly thrive.

The soil must contain a large amount of nutritive substances in a soluble state, so that the plant may be able to assimilate them, else they are not reached by the activity of the roots, and as moreover the tobacco plant attains its perfect development within a short space of time it is requisite that the means of nourishment should be afforded to it in adequate abundance.

Hence the amount of humus contained in the soil is of special importance for the thriving of the tobacco plant.

As humus originates from vegetable and animal remains, and hence affords not merely carbonic acid and ammonia but also potash, lime, sulphur, phosphorus, &c., it is specially fitted for the nutrition of plants. The mild humus which is produced by the process of decay in organic matters under the influence of heat and moisture along with free contact with the atmosphere, and which contains almost all these substances in a soluble form, has a particularly favourable effect on the growth of the plant.

On the other hand, acid humus, which is usually found in wet ground, is unsuited for the growth of the tobacco. The soil containing it may, however, be rendered fertile by incorporating with it marl, lime, or ashes, or by burning the plants growing on it, as the acid will combine with these substances and form a mild salt.

Through the mutual action of the humus and the other ingredients of the soil upon each other its effect on the growth of the tobacco plant is increased. The humus makes a clayey soil looser, drier, and warmer, a sandy soil more tenacious, as well as moister, and milder. On the other hand, clay fixes the humus, and prevents its rapid decomposition; sand decomposes it quickly, and deprives it of moisture; while lime promotes its decomposition, fixes its acids, and makes it soluble.

---

* Wien, 1866. Verlag der Central direction der K. K. Tabak fabriken und Einlos Ämter.

Through its property of absorbing heat and vapour from the atmosphere the humus keeps the soil warm and moist, and thus in the condition on which the health and vigour of the tobacco plant essentially depends.

Hence it is apparent that the tobacco plant always requires a soil rich in humus.

When however there is a superabundance of humus present in the soil, so that it becomes too loose and porous, the plants grow too luxuriantly, become sickly, or fall down, in consequence of having no firm hold on the soil.

In a loose but humous sandy soil, if the subsoil be not firm, the tobacco plant only develops a weakly stem and small leaves. Still, as such a soil is very warm, the sap is well elaborated, and the product is generally much more aromatic than in other loamy kinds of soil. In lime and marl, which attract heat and retain it for a tolerably long time, the tobacco grows much more luxuriantly, and is distinguished at the same time for its fine aroma.

Hence the culture of garden leaves which are used for the finer smoking tobacco s chiefly carried on in sandy, chalky, or marly soils rich in humus.

Small stones in the ground promote its warmth, and a soil characterized by their occurrence may then be admirably adapted for the growth of garden leaves. The more clay the soil contains the cooler it is, and the more the quality of the garden leaves deteriorates. For instance, if seeds of the garden leaves from the sandy environs of Debrö are sown in the plantations of Csanad or Békés, a product will be obtained, as large experience has amply verified, showing the characteristic properties of Szegedin tobaccos, and different in every respect from the original Debröer tobaccos.

If a sandy soil rich in humus has a loamy subsoil which protects it from drying up it is specially adapted for tobacco culture.

On such a soil cigar leaves of extraordinary quality are obtained, as for instance in the environs of Syulok. There, by clearing up the woods, a more or less sandy soil which is exceedingly rich in mould is rendered available. Such soils can be employed with advantage in the culture of tobacco for two or three years consecutively without manuring.

A loamy sandy soil with a sufficiently loamy subsoil, and so situated that it cannot become boggy, produces abundant harvests of firm set, elastic, fine ribbed leaves, which usually also trim well, and with proper management afford a rich return in cigar covers. But if the subsoil is filtering, or the soil poor in humus, the product is meagre, and is then more fit for use as cut tobacco.

A loamy soil, if the situation and subsoil are good, produces firm elastic strong leaves, the middle ones furnishing good cigar covers, and the upper ones twist and snuff.

The best and richest tobaccos, to be used as snuff, are obtained by skilful culture from soils rich in humus with a lime or marly clay subsoil which does not become boggy.

Moorland and marshy ground, which usually contain more or less acid humus, produce indeed large leaves, but these are generally loose in tissue, light coloured, of no great weight, burning with difficulty, and thence only available for ordinary cut or twist tobaccos. After frequent culture and diligent draining the product indeed is essentially improved, it assumes in part the character of tobaccos employed as snuff, but the difficulty in burning never diminishes. This defect is chiefly found in leaves grown in low lying spots, especially in wet years, when the roots of the plants sometimes remain for a considerable time surrounded by stagnant water.

Newly cut lands generally yield a high return, but the leaves are usually coarse, and thick ribbed.

When tobacco is raised uninterruptedly on the same soil it is observed that in spite of manuring and of the most industrious tillage, the plants readily sicken, while the return is uncertain, nay, is often entirely wanting.

The different kinds of soil are in this respect dissimilar.

All sorts of soil in which the process of decomposition proceeds rapidly, as with sandy, marly, and chalky ground, may be employed with advantage for the culture of tobacco during several years by a suitable amount of manuring.

The loamy sandy soils may also afford several good tobacco harvests, and the produce of the second and third years is usually finer in quality than that of the first. By longer occupation, however, failure may easily occur, especially if the decomposition of the vegetable substance is not going on favourably.

This is still more the case with loamy soils, which can only be employed with security in the culture of tobacco for two years.

When ground of this description has been employed several times consecutively for raising tobacco it may be observed, on the occurrence of a failure, that the plants become rusty and dwarfed, and that their roots spread out on the surface only, as if they sought to avoid the deeper strata. Round the points and fibres of the roots may also be observed small globules of a gelatinous slime adhering closely to these parts.

These observations lead to the supposition that the plant rejects certain useless substances which were contained in the vital sap, and which with the descending current attain the points of the roots and there secrete, accumulating in the earth around them substances which, mingling with the nutriment of the plant, render it unfit for its purpose. The fact that the rust and dwarfing of the plant, in the cases referred to, occur after refreshing showers, which usually stimulate the roots to greater activity, tends strongly to confirm this view of the matter.

If this idea is correct, the appearance of the orobranche (a sort of chokeweed) on ground which has often been used for raising tobacco is explained by the supposition that the suppurations of the root favour the development of this extremely injurious parasite. The orobranche sometimes, under the above-mentioned circumstances, appears in overwhelming quantities, adheres firmly to the roots of the tobacco plant, grows along with them, and does great damage.

This evil can only be remedied by change of soil.

Of Materials for Manure.

From the fact that the tobacco plant requires a speedy and complete development, it follows that the soil must either be naturally rich in humus, or must be strongly manured. The effect of the manure is always better when it is already decomposed, rapidly soluble, and not heating, for the success depends principally on the rapid growth, and thus on the rapid nourishment of the plant; if, however, the manure be heating, blighting of the root, rust, and failure ensue, especially in light soils.

If the tobacco be raised for several successive years on the same spot, the soil must be frequently supplied with manure, and, in this case, only rotted manure can be employed.

The rich cow dung should, where possible, be stored up before winter, spread equally and flat, so that it may not lose its power, and that its decomposition may be properly accomplished by the time for planting out.

Cow dung is the manure best fitted for tobacco plants, as, with a suitable soil, it produces rich, fine, elastic leaves; equally good is the effect of sheep-dung when this is incorporated with the soil in moderate quantity in early spring.

Horse dung is only suited for heavy soils, and must be very intimately mixed with the earth, otherwise it comes in contact in large lumps with the roots, in which case it is dangerous, as tending to produce mildew.

Excrements, from their great solubility, act very rapidly and powerfully, but cannot be employed in great quantity, and only when mixed with earth in what is called poudrette.

Birds dung and guano used alone are poisonous and destructive. Through the large quantity of ammonia which they contain they are very active when diluted with water, or ground small and mixed carefully with earth.

Liquid manure is also of great value, on account of its abundant supply of ammonia and potash; it is diluted with about six times the quantity of water and poured round about the growing plant, with most effect in moist weather.

Whether, as is frequently said, tobacco receives any peculiar smell or taste from horse, sheep, or swine dung has scarcely been satisfactorily proved.

From green manuring tobacco receives a mild nourishment, and this kind of manuring is consequently of high practical value.

In reference to the strength of manuring, it must be borne in mind that the smaller the amount of organic substances in the soil the stronger is the manuring required. A weak manuring of a soil already exhausted, and a strong manuring where there is sufficient power without it are alike faulty. The quantity of manure employed should be sufficient for the perfect development of the plant, in order to secure a rich return by vigorous growth.

On the quality of the manure, as before stated, its strength also depends. Sheep and horse dung work quicker and more powerfully than other kinds of stable manure, but their effect is less durable, on which account, when using these first, less must be put on at a time, but the operation must be oftener repeated.

With fresh dung the manuring must be stronger than with rotted.

Hard, cold, moist soils require to be more strongly manured, but not so frequently as active, sandy, dry or chalky soils. For these last a weak but frequently repeated manuring is suitable.

Sloping ground should be more strongly manured at the top than below, because the rain will wash down some portion of the manure from the top to the bottom.

Manuring may be classified as strong, medium, and weak. Of cow dung, not too much fermented, 30 loads of 10 cwt. each, or 1½ tons per acre, is reckoned strong manuring, 22 loads of 10 cwt. per acre medium manuring, and 15 loads of 10 cwt. a weak manuring.

In what cases stronger or weaker manuring is to be employed has already been explained.

Of other sorts of manure the following may yet be mentioned:—

Blood acts well as a manure, from the rapidity of its decomposition and the abundance of nitrogen which it contains. It must either be diluted with a quantity of water, or mixed with earth.

Bone dust does not much further the growth of the tobacco plant, because, even when set free by sulphur or oxalic acid, it is yet too slow in decomposing.

Burnt lime draws carbonic acid from the atmosphere, and gives it out to the plants. It moreover promotes the decomposition of organic substances present in the soil, making them fit to afford nutriment. Acid humus it makes mild and fertilizing. On account of this property and its own value as a manure, burnt carbonate of lime is an excellent manure for the tobacco plant. It is, however, only available on strong loam and clay soils and ground freshly brought under tillage; meagre, sandy soils, and also such as naturally contain abundance of lime, are unsuited for it.

Manuring with lime must not be frequently repeated, or the soil would become too much weakened. The lime is reduced to a fine powder either by sprinkling it with water, covering it with earth or sods, or turning it over. This powder is strewed on the ground before the last ploughing, and well harrowed in, dry weather being the most favourable for the purpose. Another very beneficial effect of the burnt lime is the destruction of vermin present in the soil, as these are very injurious to the tobacco plant.

To an acre of surface from 10 to 30 pecks are sufficient, according to the special character of the soil.

Less powerful in its effects, yet still beneficial to the tobacco plant is manuring with gypsum.

Gypsum draws ammonia from the atmosphere, and combines with it, so that with every shower that falls some of the ammonia is separated and dissolved by the water. Moreover the soil is enriched by the gypsum in lime and sulphuric acid. The gypsum is strewed in moderate quantity on the surface of the ground round the growing plants, and the best time for this is in the evening when it threatens rain.

Marl is only suited for such soils as are poor in lime, and where consequently the increase of this substance is required.

Ashes consist of the mineral ingredients which plants have drawn from the soil, and as a portion of these are soluble in water, they are available as a manure, but their influence on the growth of the tobacco plant is very trifling.

Compost manure, as being a well prepared sort of nutriment of very rich earth, invigorates the soil in a considerable degree, and may essentially promote the growth of the tobacco plant.

The quantity required for an extensive plantation is however rarely procurable, and its employment is chiefly limited to the cultivation of beds.

Compost is prepared by alternating layers of common earth, sods, and mire or street-mud, from 4 to 6 inches thick, first with vegetable and animal remains, sawdust, straw, sweepings and animal droppings, next with decomposed lime, marl, and ashes to the height of about 5 inches; then, from time to time, the heap should be wetted with liquid manure or dirty water, and, when fermentation has taken place, it should be turned over for the cooling of the interior, this turning over to be repeated from month to month.

The before-mentioned substances are sometimes mixed in a different way by being thrown at once into the heap with the requisite quantity of earth, strewed in layers with lime or ashes, and otherwise treated as above described.

Tobacco as a Rotation Crop.

When tobacco is planted on the three years system of rotations, it is best to set it in fresh manured fallow ground, this crop to be succeeded by a winter one, as wheat or rye, and in the third year by a summer one, as barley, oats, &c.

The fallow ground, however, should always be manured; but in Lower Hungary, where the soil is mostly very rich, this need scarcely be done oftener than once in six or nine years.

It is therefore attempted to improve the soil by means of hoeing-plants, and the following rotation is observed :—

    1, Hoeing-plants (maize, tobacco); 2, Wheat; 3, Fodder plants, and afterwards, stubble pasture and summer fallow.

On less fertile soils where the land lies entirely fallow only such spots should be chosen for the culture of tobacco as can be subjected to a stronger manuring. Spots of this kind are the more easily obtained, as in such districts the culture of tobacco is only carried on on a smaller scale.

In the rotation of crops it is to be observed that tobacco requires a strength of manure corresponding to the condition of the soil. The second year after manuring is the best suited for planting, especially if cow dung be predominant. The following are instances of the rotation of crops followed in Lower Hungary :—

A. 1. In strong manure, plants for fodder (vetches, oats, green crops, &c.)
    2. Tobacco and other hoed-plants.
    3. Wheat.
    4. Barley, oats, &c.
    5. Maize.
    6. Wheat or rye.

B. 1. In manure, tobacco.
    2. Wheat.
    3. Foddering plants or summer pasture, then fallow.
    4. Rape.
    5. Maize.
    6. Wheat or rye.

C. 1. Tobacco, in manure.
    2. Wheat.
    3. Summer crop, for instance, barley with clover.
    4. Rape.
    5. Winter crop.
    6. Oats.

D. 1. Fodder-plants, in manure.
    2. Tobacco.
    3. Wheat.
    4. Summer pasture, with clover.
    5. Clover.
    6. Winter crop.

In less fertile soils the vigour must evidently be sustained by a small amount of manuring within the term of rotation.

The exigencies of agriculture and the nature of the soil may modify in various ways the rotation here given, the foregoing examples are only meant to point out some of the objects most suitable for cultivation in the same spot as the tobacco, and which may most conduce to its success.

Preparation of the Soil for the Cultivation of the Tobacco Plant.

Thorough loosening and mixture of the soil with manure will greatly contribute to the growth of the plant, for through the loosening of the soil the ground is opened to the influence of air, heat, and water, and the decomposition of its ingredients as well as that of the incorporated manures is favoured, and thence their transformation into food for plants; further, injurious weeds are destroyed, and the plants under cultivation are enabled to spread out their roots sufficiently on all sides.

Stiff soils require a more frequent and special loosening than sandy soils. They must, therefore, in the first place, be deeply ploughed in autumn. If the tobacco be manured, and fresh cow dung be employed, it is most suitable, as has been already said, to lay this down in autumn. Should this however not be possible, or should putrid stable dung be more at command, it must be collected in the winter or at latest in the beginning of spring, immediately spread equally, and as early as possible covered over flatly. This process must also be observed on light sandy soils or those containing much lime.

Before setting the plants the ground must finally be ploughed deeply and evenly, and then carefully harrowed, in order to diminish the clods of earth, and to give it an even, regular, surface. Heavy hard ground should therefore be ploughed over thrice, light soil twice.

The use of the roller is beneficial under all circumstances, especially as it makes the surface even, lessens clods, prevents the formation of hollow spaces between, presses the ground on the surface close, and thus guards against too rapid drying. On this last account it is most serviceable in light soils.

Position and Protection of the Tobacco Field.

Lastly, the situation of the soil chosen is important to the success of the tobacco plant. The produce is always finer and better when the plants have abundantly enjoyed the influence of light and heat.

When the ordinary surface of level lowland does not induce too much moisture it is well fitted for the culture of tobacco, and a level situation is always better than a steep declivity.

A gentle slope towards the sun however produces the best effects, provided the soil is sufficiently tenacious not too easily to suffer drought. The southern slope is then the most desirable, the northern the worst. When the declivities are considerable, however, the plants suffer much from the rushing down of the rain water.

Broad valleys lying between hills are usually very fertile and sheltered, and are thence well fitted for the cultivation of tobacco if the climate is also suitable.

If the field devoted to tobacco culture could, along with the conditions of favourable situation and protection from flooding, be found in the neighbourhood of a river, brook, or pond, so that soft water could be procured at no great distance for watering the plants, the arrangement would be very desirable.

The use of soft water for this purpose is always to be preferred to that of spring water. Hard spring water often contains so much mineral matter in solution that it possesses little dissolving power to furnish the substances necessary for the nourishment of the plant.

If the spring water contains abundance of soda and saltpetre, as is frequently the case in Lower Hungary, it should only be used for watering the plants, in the last extremity.

As the tobacco leaves are extremely deteriorated by being torn by blasts of wind, their protection against these must be provided for as much as possible. For this purpose hedges, especially on the north and west sides, which break the wind without barring its entrance, are much to be commended, especially as they further retain the warmth of the soil and moderate its drying. High hedges and walls are very useful for the same reasons, as also because they keep animals off the plantation.

If however no fence whatever can be afforded to the field, it is well to plant round it tall plants, as maize, giant-beans, sunflowers, &c., and if the surface be very extensive, strips of the same in suitable directions should be planted in the interior, in order to moderate the injurious effect of the wind.

TOBACCO SEED BEDS.

Laying out of the Beds.

The tobacco seed is first sown in plant-beds, and only when, after from 4 to 8 weeks, the plants have attained a certain size, are they transplanted into the fields.

The beds should enjoy the influence of the sun throughout the whole day, and must be protected against rough weather and blasts. A spot lying on the south side of a building, a hedge, or a shrubbery with sufficient inclination for the water to run off is the best, and on the contrary a hollow in which water collects during heavy rain is the worst for the purpose.

Beds for the tobacco plants are classified as cold, warm, and hot beds. A cold bed is one in which the planter turns over a good free garden soil, makes it smooth with the rake, and strews the seed on it. It is good to add a little rather putrid dung to this earth. Such beds are not suited for developing the heat necessary for the rapid growth of the plant, which consequently is very late in attaining the size requisite for transplantation. It would therefore be desirable, in order to attain a moderate degree of warmth in the ground, to put aside the earth turned up in the first place, fill the space thus made with a layer from 4 to 6 inches thick of fresh horse-dung, press this firmly down, and on the top of it place the garden mould previously set aside.

Stable dung is generally employed for the development of heat. Fresh horse dung in this case deserves the preference. This last soon rots or ferments, by which heat is engendered and communicated from the layer of manure to the earth resting on it. The thickness of the layer of dung depends on whether or not the bed is so constructed as to retain heat. The more the heat can escape sideways, the more wind and rain have entrance, the stronger should be the layer of dung.

The best, most fertile, and light garden mould should be chosen for the formation of the beds, in order that adequate nutriment may be afforded to the young plants. The earth should moreover be as free as possible from the seeds of weeds (which of course must have been previously seen to), and must be very loose, and without clods, stones, or fibres of roots; these requisites are most readily attained by passing the mould through a wire sieve.

The best soil for tobacco beds is prepared in compost heaps of the previous year, as formerly described in the chapter on manures.

The thickness of the layer of earth on the bed should be about 6 inches.

The preparation of the hotbed is as follows:—

The space designed for the bed is surrounded by a fence of hurdles or boards about 2 feet in height. In the inside of this enclosure fresh horse dung is laid, firmly trodden down in layers and filled up till the dung reaches the height of from 1 foot to $1\frac{1}{2}$. The dung now begins to be strongly heated, as is seen by its steam. The steaming commenced, the prepared earth is laid on to the height of 6 inches, distributed as evenly as possible, and smoothed with the rake; immediately afterwards the sowing of the seed may be executed. In the daytime, and in warm weather the bed may be protected with brushwood, but in the evenings, and with cold unfavourable weather, they should be protected from injury by a covering of mats.

As the dung decays the bed sinks further down, so that the surrounding fence projects more and more above the surface of the earth, and thus is convenient for a support to the covering of mats, straw, or rushes, under which a few cross beams are placed to keep them always at some distance from the soil, so that they may not interfere with the germination of the plant.

Instead of placing the dung on the surface of the ground, it is a better plan to clear the earth from the bed, to the depth of about a foot, and then to make the fence only 1 foot high, following in other respects the foregoing directions.

Such beds for tobacco plants require no great outlay, so that the poorest planter may employ them, if he does not shun the small amount of labour required for their construction.

For increasing the forcing power of hotbeds, arrangements may be made for covering them with a frame of glass or paper, in order to increase the action of the sun's rays and to prevent the immediate escape of the vapour evolved.

When a framework of wood, at least 6 inches deep and provided with glass or paper fittings, is placed upon the hotbed before described, a forcing bed of great simplicity is produced, which, in comparison with the uncovered bed, affords incomparably better results.

Such beds have, however, many defects. If the outside is not protected by a frame, warmth and nutriment are lost by the free entrance of the wind and the washing away of the dung by the rain in wet weather. This is especially the case where the beds have no surrounding fence. In the contrary case, the space between the surface of the soil and the glass resting on the framework is regulated by the breadth of the boards, and hence is commonly too low; the warmth generated within and the rays of the sun without combine to produce a strong heat, which may be very injurious to the plants, if the beds be not carefully aired by raising the glass.

This evil will be most thoroughly remedied by surrounding the bed with a framework, of which the side fronting the south should be a plank's breadth in height, and that at the back $2\frac{1}{2}$ to 3 planks breadths. The boards forming the walls are fastened by grooves into stakes stuck into the ground, at a distance of 3 to 4 feet from each other. The stakes standing opposite are joined at the top by cross-ledges. By this arrangement the covering glass* or paper frames receive an oblique direction, inclining towards the south.

The paper intended for covering the frames is first moderately damped with a sponge, sheet by sheet, then laid one on the top of the other and wrapped in a cloth folded several times, and also somewhat wetted so that the moisture may be equally distributed. After some hours, if the sheets of paper no longer show moisture but feel damp, the fixing of them on the frames may proceed; for this purpose flour paste should be used, laying it on with a pencil, and thus ensuring the painting of the frame edges. The frames thus covered with paper are set out to dry either in the open air or in some airy place, after which they are again painted with boiled linseed oil and dried afresh. The frames when finished resist rain, and serve their purpose almost better than glass frames. These last when the sun strikes on them immediately create a great heat, whilst under frames of paper the heat is developed much more slowly. If the weather be chill the glass frames alone cannot protect the beds from cold, whilst paper frames retain the heat within them, paper being a bad conductor.

In order to procure sufficient light in the beds for the development of the plants a white paper is chosen of equal and fine texture, but not too thick, and oil of pure and good quality. If by any accident a hole is made in the paper it must be mended by pasting on a new piece, which must also be oiled. After a night of severe frost too, when the oil freezes, the frames should be slightly oiled. A single covering of paper may last for several years if kept free from dirt, but the process of oiling must be repeated every year.

Size of Seed Beds.

The size of the beds is dependent on the extent of surface to be devoted to the cultivation of tobacco, the species to be reared, and the relative space required by each individual plant for its development without injury to those around.

The number of plants set in an acre of surface varies according to the mode of tillage and the kind of tobacco.

In the cultivation of heavy leaves used for snuff, ordinary cut tobacco, and cigar covers, the surface of an acre will require from 10,000 to 14,000 plants to set it; if, however, it is designed for the production of lighter cut-tobacco (garden leaves), the plants are set considerably closer, and an acre of surface will require from 30,000 to 50,000 of them.

Frequently a large portion of the plants set perish, in consequence of unfavourable circumstances of the weather, worms eating them up, &c., and the empty spaces thus left have to be immediately filled up. Further, if every individual plant set is not vigorous, all such, whether ill-rooted, sickly, or crooked stemmed, must be removed.

Lastly, it must be borne in mind that the plants in a bed do not all grow with equal rapidity, and that many do not attain the size required for transplanting until after the time when this must necessarily take place.

It is therefore advisable to provide the beds with a considerably larger quantity of plants than will be required for a single transplantation. In order to be secure under every circumstance, a good planter will provide himself with at least twice, or even three times the quantity of plants required, and the more so as the possible superfluity will always find a ready purchaser, whilst a deficiency is certain to prove a loss.

In order that the tobacco plants may develop perfectly in the beds, sufficient space must be allowed to their roots and leaves, so that they do not hinder each other's growth. If the seed, as commonly happens, is too thickly strewn, one plant stands very near another, the roots cannot spread out sufficiently, or they become entangled with those of their neighbours, and are thus injured when removed for transplantation, even if several of them be not, unavoidably, pulled up at once.

Besides, when closely set, the plants lack sufficient space for the growth of their leaves and shoot up in height, thus producing thin stems and small leaves.

---

* There is one covering frame for every interval between each two cross ledges. To prevent, in heavy showers of rain, the water falling on the cross ledges from penetrating into the interior, they must be provided with a groove running along the middle, so that the water may easily run off.

That such tall-stemmed, thin, ill-rooted plants are not well fitted for transplantation must be obvious to every one. A further disadvantage of planting too thickly becomes apparent after the first or second removal of the stronger plants. So long as the plants stand close in the bed the one supports the other, but when there are many withdrawn the remainder are apt to become bent or even to fall over.

This is especially the case with those of a thin long stem, which is incapable of supporting the crown of the plant in an upright position; such plants especially are particularly liable to be bent, under the indispensable operation of watering, or when wind finds entrance, and thus the stem becomes so crooked as to be unfit for transplantation.

If the seed be so scattered that not more than four plants stand on every square inch of the bed, the evil of thick sowing will be perfectly avoided, and vigorous, well-rooted plants will be obtained.

The bed is best fitted for its purpose when of the breadth of four feet. With a decrease in breadth its suitability will diminish. To give it a greater breadth than four feet would entail many inconveniences in the tending of the plants.

Hence, in order to produce plants enough to sow an acre, a bed four feet in breadth will require to have a length varying as follows:—

For heavy leaves, suited for snuffs and cigar covers, or otherwise intended principally for the production of large leaves - 18 to 23 feet.
For garden leaves of large size - 23 to 30 feet.
For garden leaves of small size - 30 to 45 feet.

Quantity of Seed for a Bed of certain Dimensions.

If the seed destined for the raising of tobacco consisted entirely of perfectly germinating grains, and if in sowing it one could ensure that on each square inch four little seeds should fall, the sowing of an acre of ground with the small-leafed tobacco would scarcely require half an ounce of seed; but as plants in their earliest stages are least able to withstand unfavourable external influences, and are most easily and frequently deprived of life, we must bear the fact in mind when sowing, as also that many grains have already lost the power of germination; that one portion of the germinated seed dies in its earliest youth; that another is injured by insects; and, lastly, that the sowing can never be so accurately performed that there will not be spots where some must be drawn out by hand, and, therefore, that a suitable addition must be made to the quantity of seed which would otherwise be required, in order that the bed may be equally set with plants, and no vacant spaces left.

Cold beds require more seed than warm open ones, and these last again more than forcing beds with glass or paper frames.

The quantity of seed required according to the number of plants to be set per acre, and the dimensions of the beds as above given, is as follows:—

A. Of fresh seed of the preceding year's growth:—	In cold Beds. ozs.	In warm open Beds. ozs.	In hot Beds. ozs.
(a.) For heavy leaves, and especially for large leaves	1 to 1¼	¾ to 1	½ to ¾
(b.) For garden leaves of the larger sort	1¼ to 1½	1¼ to 1⅜	1 to 1¼
(c.) For garden leaves of the small sort	1⅝ to 2⅜	1½ to 2	1¼ to 1⅚
B. Of seed two to three years' old:—			
(a.) For leaves of large size	1¼ to 1½	1 to 1¼	¾ to 1
(b.) For garden leaves of large size	1⅔ to 2¼	1½ to 1⅞	1¼ to 1½
(c.) For garden leaves of small size	2⅜ to 3¼	2¼ to 3	1⅗ to 2¼

The Sowing of the Tobacco Seed.

When the bed for the plants has been prepared in the manner described, and there is no longer cause to dread the occurrence of severe frosts, the sowing of the seed may at once be proceeded with.

This usually takes place in the latter half of the month of March or the beginning of April. Hot beds may be sown much earlier than cold beds.

As the setting of the plants where large extents of surface are to be covered always demands much time, the planter will do well to prepare both warm and cold beds, so as to have a constant supply of vigorous young seedlings.

The seed may either be sown dry, or it may be moistened beforehand to promote germination. For this purpose the seed is put in a linen or woollen bag, frequently moistened with water, and kept moderately warm in the neighbourhood of a stove, but care must be taken to repeat the process moistening so as to prevent its drying up.

In this manner also the germinating power of doubtful seed may be tried, only a very small quantity of it of course being required for the purpose. A healthy seed germinates according to its age in from four to eight days; if germination does not by that time take place, the parcel of seed from which the sample was taken cannot be used with confidence.

The seed designed for sowing should only swell, but not burst the shell so as to expose the germ to view, because in this case it might easily receive injury in the manipulation of sowing.

The sowing should only take place in calm weather; if wind or rain occurs it must be delayed.

Before the sowing, the surface of the ground should be stirred with a rake, but only to the depth of about an inch.

The grains of seed are too small to render it possible to strew them singly with accuracy, and the difficulty of a uniform sowing is increased by the fact that the ground on which the seed is strewn is dark, and thus prevents the spots where it has fallen from being very apparent. The seed is therefore mixed with perhaps ten times the quantity of sifted wood ashes or fine dry sand, by which means the process of sowing is exceedingly facilitated. The sand or ashes chosen for this purpose should be as white as possible, in order to point out very conspicuously the spots sown. Whilst the mixture is held on a plate or other vessel in one hand the sowing is effected with the other.

It is a good plan not to proceed with the sowing of the whole bed at once, but to divide both it and the seed into a number of small equal portions, because in this way the sowing can be more easily controlled, so as not to have too much or too little seed in any spot. The opinion that the operation is more uniformly effected by means of a sieve than with the hand is not tenable, since by an accidental increase of force in shaking the sieve more seed will fall through than with gentler motion, while the specific gravity of the ingredients will make the heaviest when thus shaken come to the bottom of the sieve, and thus fall first to the ground.

On the other hand, it is a good plan immediately after the conclusion of the sowing to cover the ground with a layer of fine garden mould or perfectly rotten manure not higher than half an inch, and to effect this by means of a sieve, because, on account of the germinating process already referred to, this covering should be light and free.

The tending of the Beds.

The care of the beds demands much attention, because the tobacco plant in its earlier stages is extemely delicate.

After the completion of the sowing the bed is duly watered. A loose humous soil absorbs much water, but a loamy one less, and the amount of watering must therefore be regulated by the nature of the soil, and must never be carried so far that the water remains standing on the surface. In watering, care must be taken to avoid too great a rush of the fluid at once, in order that the earth may not be washed away. For this purpose a watering can with a finely drilled rose is used. Poor planters dip a wisp of straw in water, and thus besprinkle the bed, or they let the water fall through a sieve.

When the plants have attained a certain size, the water must be sprinkled in very small drops for fear of bending their stems.

The quality of the water used for the nutriment of the plant is of great importance. This subject has already been copiously discussed in the chapter on the position of tobacco fields, and it only remains to be noticed that when water is too cold it is injurious to vegetation, and therefore before being used for this purpose it should be warmed by being allowed to stand, or by having heated stones thrown into it. Rain water collected from the eaves produces the best effect. The natural moistening of the bed is very advantageous, but only when the rain is gentle and warm; in stormy or cold weather the frames of the hotbeds should never be raised, and the common open beds should, as far as possible, have a good fence of straw or rush mats, boards, &c.

After the first watering of the beds the frames are put on by way of second covering, and usually are not raised again till the second or third day for the repetition of the watering and for air.

The further watering of the beds is then to take place only when the soil is evidently dry, and no more water must be given than is requisite for moistening the soil and being easily absorbed.

After each watering the frames should at once be closed in order to remedy as speedily as possible the cooling which has taken place.

The heat in the tobacco bed must not exceed a certain height, and should never be so great that on raising the frame it is strongly felt. The periods at which the frames should be opened are shown by the drops adhering to the inside of the glass or paper in consequence of the great evaporation going on. So long as the heat within is moderate these drops are small, but if they are seen to increase considerably in number and size, the raising of the frames must not be delayed. In order to effect this raising more easily the frames must be provided with rings, and indented supports of wood should be constantly at hand, so that the frame may be kept open at any height desired. In windy weather the beds are sufficiently aired, and the raising of the frames is then superfluous. In order however to prevent them from being shaken and injured by the wind, they may be kept fast by means of wooden buttons, or still better by iron hooks and clasps. In fine warm weather the beds are usually watered once a day. The best time for the operation is 9 a.m., the worst towards night. In cold weather omit the watering.

The beds must be carefully protected against cold during the night. A hotbed with a higher wall to the back and well fitting frames preserves the requisite degree of heat even in a cold night. It is well, however, to cover the frames with mats, especially when frost is feared, because the oil on the paper freezes, which lessens its efficacy.

If severer frosts occur whilst the plants are in the bed it will be well to surround the side walls on the outside with dung, so that the injurious cold may not be able to penetrate the joints of the boards. A hotbed with horizontal frame should always be protected from the cold during the night, because the space allowed for air between the frame and the soil is too small. If the open beds are early set, they cannot be sufficiently protected against cold by straw, mats, boards, &c.

If the foregoing precautions and conditions be duly observed, the tobacco plant will make its appearance above ground in from five to ten days, according to the kind of bed, and the quality of the seed as regards its age, and whether sown dry or after being soaked. The care of the bed during this time requires no change. If in the course of 10 or 14 days after the sowing a corresponding number of plants does not appear, the empty or thin spots should be sown over again with moistened seed.

The young plants suffer if the bed be too much watered, overheated, or too much cooled. In every case such an injury to the plants is early noticeable, because they either become withered, or change their beautiful bright green colour for yellow or black.

Those who have charge of the beds will easily judge what has injured the plant.

If the watering seems to have been overdone it must be discontinued for some days, and in fine weather, but not when the sun is too hot, the beds should be kept open. If cold has accidentally found entrance the injury cannot be quickly remedied. It is a good plan to water sickly plants with water in which a little poultry or cow dung has been dissolved, or with greatly diluted urine. But if the plants are burnt up with overheating or are frost bitten, the best thing to be done is to rake over the bed and sow it anew.

The tobacco plants are frequently injured by vermin while yet in the bed, but close beds afford most protection against this. Springtails and other insects are kept out of the bed by the use of ashes instead of sand in the sowing. If vermin are still to be found in it, they may be driven away by sprinkling it with water in which boiled tobacco or powdered chalk has been stirred, or with greatly diluted urine.

The mole cricket digs long passages near the surface of the ground, bites all the roots of the plants that come in its way, and thus when numerous is extremely injurious. Their presence betrays itself by the withering and death of the plants, and by striped elevations of the ground which explain this token. These elevations must be followed up till the insect and its nest are discovered, the latter containing usually myriads of eggs. These insects are driven away by ants, and these again by frequent waterings with hot water, and repeated destructions of their dwellings. Garlic placed in the galleries of the mole crickets will also drive them away.

For the mole it will be necessary to set traps and try to catch him. Frogs cannot be suffered on the beds because the leaves which they have rendered slimy immediately rot. If snails appear, the beds must be strewed with chopped straw or chaff, over which they cannot crawl.

With the progressive development of the plant the hotbeds are opened with increasing frequency in order to refresh and invigorate the tender plants, and to accustom them to the external atmosphere. Latterly therefore the beds are kept shut only when the heat of the sun is intense, during stormy weather, in the evenings, or on a cold day. At last the more advanced plants, when there is no fear of hoar frost, remain uncovered even during the night.

The opinion that plants reared in hot beds are less able to resist the influence of bad weather than those grown in cold beds rests principally on prejudice, for by the regular attention to the beds recommended above they are gradually and perfectly hardened, so that they can be transferred to the field several weeks earlier than the others, which is always of very great importance.

Along with the germinating tobacco plants, weeds make their appearance in greater or less number, according to their greater or less prevalence in the earth employed in the formation of the beds. Weeds usurp both space and nourishment, and are thus extremely prejudicial to the growth of the tobacco plant.

When the weeds have grown large enough to be grasped between the fingers without injury to the tobacco plant they are plucked up. In weeding one must beware of drawing out fungi, for these increase much when so treated. The weeds plucked up must be entirely removed, for if a single little plant be left lying loose on the surface of the bed it will immediately rot and infect all around it. If patches of decaying plants are discovered, they must be removed forthwith, because the infection spreads rapidly.

Before and after weeding the beds must be watered, before the weeding in order to make the earth part more easily from the roots, and after the weeding in order that the earth which has been too much loosened may be put back again over the plants which remain in the bed.

Transplantation of Thinnings or Pricking Out.

If the tobacco plants stand too close in the bed they mutually exercise the same effect as weeds. In spots where this occurs the superfluous numbers must be pulled up, and this must be done when the little leaf has grown large enough to be grasped by the fingers, and when the stalk can bear the tug without tearing.

The small and tender tobacco plants thinned out of the beds are not thrown away, but planted in ordinary garden beds. For this purpose spots are chosen either in the garden or the field where the soil is of the best quality, and these are made as sheltered as possible, by putting heaps of earth or fences about them, if it be necessary. The earth is turned over, rendered fine, smoothed with a rake, and divided into beds 3 feet in breadth. In these beds the little plants are placed, a hole being made with the finger or a dibble, and the earth pressed down on the root. The distance of each plant from the next should be an inch. To save space it is well to set the plants in rows, and the distance between the rows should be also an inch; it should, moreover, be observed, that each plant does not stand opposite that in the next row, but midway between, so that those in alternate rows may correspond, as is shown in the accompanying figure.

. . . .
. . . .
. . . .

The transplanted beds are immediately watered with a watering-can, and further attended to in the same way as the sown beds. Of course they must also, especially at first, be protected from the heat of the sun, and constantly from frost by being covered over with mats.

On a bed six feet square about 5,000 plants may be set in this way.

The plants thus transferred to beds in the open air, if treated in the manner indicated, grow rapidly; having sufficient space, the stem root grows strong, does not shoot to a great height, the stalk becomes thick and vigorous, and the leaves not so tender as those of the plants remaining in the seed bed, but much firmer and more luxuriant; such plants also resist the influence of unfavourable weather much more easily.

When the plants thus treated are afterwards set in the open fields, a considerable quantity of the earth in which they have grown remains attached to their roots, and thus no material shock is given to their vigour by the change of soil; they do indeed fade a little, but the leaves at the heart retain their freshness, and are developed soon after those of plants which have been set in the open field a good while previously.

The plants in question are especially fitted to be seed plants from their luxuriousness, while from their rapid growth they are fitted to supply the place of those that are destroyed by worms, or by other unfortunate casualties. When vacant spots in the field are filled up from the seed beds the new comers, not being able to overtake their neighbours in growth, are soon deprived by them of the influence of the light so necessary to their development, and thus always remain sickly, and seldom ripen, whilst the early transplanted roots when employed for this purpose, as they possess greater vigour, soon equal those around them in growth, and are thus secure from this obstacle to development.

Tobacco planters will then do well carefully to attend to the setting of the plants thinned out of the seed beds as superfluous.

PLANTING THE TOBACCO.

If proper care be bestowed on the beds the plants grow rapidly and are soon fit for transplantation. A hotbed composed of good manure and earth rich in humus furnishes plants fit for the purpose in four or five weeks; with beds of another kind six or eight weeks are required. If the beds are sown towards the end of March the transplanting may begin early in May.

A plant fit for transplantation must have a good root, the stem must not be too long or thin, and especially must not be crooked; its development must have advanced as far as the unfolding of four or five leaves, the largest of which must be at least an inch in breadth; lastly, it must be free from disease, such as blight of root or stem, and from injury, particularly to the stem or heart leaves.

All the seed sown on the same day does not produce plants of equal size at the same period; in each bed, along with plants fit for setting, a great many more will be found much less advanced in growth. This however is no disadvantage, because where there is a large extent of surface to be set the whole work cannot be done in a day, nay, frequently not in a week. Only those plants then that are full grown are withdrawn, and the rest are allowed to stand till they also have attained the desired size.

In taking plants for setting the same precautions should be used as in the removal of weeds. Before plucking up the plants the beds should be duly watered. The plants to be pulled up should be grasped by the strongest leaf with the fingers of the right hand, drawn gently till they leave the soil and carefully placed in the left hand without disturbing the earth adhering to the roots. When several plants have been pulled up, they must be laid in a basket or in some suitable vessel, where they should be covered with a moist cloth or with grass to preserve them from withering.

The removal of the plants should, if possible, be delayed in windy weather, because the plants remaining in the beds are thus apt to become crooked. No more plants should be taken up than can be set the same day. The plants, if well protected, may indeed if necessary be kept for several days, but many of them by this means are so weakened that they can scarcely recover themselves.

The plants must be set in the fields in rows at equal and measured distances, so that they may not mutually hinder each other in their development, and that access may be got to each without injury to the leaves, whilst at the same time the utmost use of the surface must not be lost sight of. Whilst on the one side the requirements for successful growth and consequently for corresponding attention must be borne in mind, on the other, we must endeavour to have the greatest possible number of plants in order to obtain a rich return of leaves. The plants must therefore not be set farther apart than is necessary for the full development of the leaves and for obtaining free access to them. It is not, however, necessary to devote so much space to each plant that it can be got at on all sides, because all the manipulations can be equally well performed when the access is only at one side. It hence follows that there is unnecessary waste of space where the plants are set equally far apart on all sides, and when their distances exceed the space required for the free development of the leaves.

Most planters observe equal distances in the arrangement of the rows, as is shown in the annexed figure.

It is evident that if one walks between the rows a and b, both these can be tended, so that there is no need to make use of the space between b and c, since the rows c and d can be attended to by means of the succeeding space.

It is therefore useless to leave a wide space between b and c for the purpose of affording access to them.

If the distances between the rows, where a passage is not necessary, be limited to what is requisite for the development of the leaves, these will be placed at unequal distances, and indeed the narrower the rows without passages the wider those with them. The accompanying figure makes this more convenient arrangement intelligible.

The advantage attained in this manner is not inconsiderable, and consists in this, that, without the slightest detriment to the plants, the number of rows is increased in the proportion of $4:5$, so that 25 per cent. more plants can be set on the same extent of surface, and consequently so many more leaves can be gathered.

A further improvement in the mode of planting consists in placing the plants in contiguous rows opposite each other, but in such a manner that each plant is opposite to one in the second row from itself, and midway between those in the one next to it, as is shown in the accompanying figure.

In this way the plants have the greatest space possible for spreading out, and at the same time protect each other better from the wind.

The distance between the plants in a row must depend on the size of the leaves. If the plants are too thickly set, they hinder each other in their development, if on the other hand they are too sparse, the ground has not been well employed.

The distance of the rows and of the plants must therefore be measured by the species of tobacco and the quality of the soil.

In raising heavy leaves for the manufacture of snuff the greatest space relatively to their size must be allowed, in order that they may enjoy as much as possible the influence of the sun, and successfully carry on the processes of vegetable growth. Tobacco for cigar covers may be set somewhat closer, because thickness is not so much aimed at in this case as fineness of texture; it is, therefore, not to be considered a fault if the leaves of neighbouring and opposite plants lap over somewhat, and thus partially moderate the influence of the sun. Plants with small leaves, as the garden leaves, are set thick. Certain kinds of these leaves as the "debrö" only attain the qualities desired when they are small; if a large size be demanded, it can only be had at the expense of the quality. The small size, therefore, requires, on the one hand, that the plants be set thick, on the other, that the soil be loose sand.

The following table shows the distance to be observed in the arrangement of the rows and of the plants in different sorts of tobacco.

Kind of Tobacco.	Distances in setting the Plants for		
	Rows		And distance of Plants in a row.
	Without passage.	With passage.	
	Inches.	Inches.	Inches.
Original Galitzian heavy leaves for snuff making	21	30	24
Hungarian leaves for snuff, large size	21	33	21
Hungarian leaves for snuff making, small size	18	30	18
Cigar leaves and common leaves, large size	18	33	18
Common leaves of middle size	15	27	15
Common leaves, small size, and also for gartenblätter of large size as in Siebenbürgen	12	24	15
Middle-sized garden leaves	10	18	12
Small-sized garden leaves	8	16	10

In order to keep the correct distance between the rows of plants during the setting, a marker should be prepared, consisting of a beam with three projecting wooden pegs on both sides, and a handle by which to guide it. Two of these pegs show the distance from one row to the other, while the distance of the second from the third is the width of the space allotted to the passage.

If this marker be drawn across the field in the direction in which the rows of plants are intended to be set three perceptible furrows or ruts are obtained. At the ends of the rows the marker is inverted, so that the prongs which before were uppermost now stick into the ground, and the outermost prong being set into the last furrow and the marker drawn backwards, two new furrows are obtained, and so on till the whole field is marked. If the marker were provided with prongs on one side only, the inequality of the distances between the rows would prevent its working alternately backwards and forwards. Only one movement would then be efficient, the reverse one merely serving to bring the instrument into position. The loss of time thus occasioned is saved by having the prongs pass through the beam and project on both sides.

In fields of considerable extent it is a good plan, to leave every 20 yards or so, a free space in the rows of plants from three to four feet wide for roads, in order to be able to have unobstructed communication at all times with every point of the plantation. If these roads are not left free the watering of the plants cannot in many places be satisfactorily executed, the labourers pass over the field at their pleasure, doing much damage by trampling down the earth, and further on in the season it will be necessary to pass for long distances between the ripe tobacco plants, frequently also causing much injury to the leaves. Besides at the time of the gathering the leaves may be spread for the time on these paths before they are taken to the drying-house, and it is therefore desirable that the rows should not be too long, as otherwise the going and coming will consume much time. According to the shape of the field, therefore, suitable paths must be traced by means of a cord stretched across, before proceeding to the setting of the plants. After this follows the furrowing of the field into several divisions by the use of the marker as already described, and then the transplantation.

This is done in the following manner:

Each labourer takes a number of plants on a plate or in a little basket and places them at his left side, so that each time he can lift a plant with his left hand carefully grasping it by the leaves. Meanwhile his right hand is occupied in boring with the dibble a perpendicular hole the exact depth of the roots, after which he presses the dibble somewhat towards himself. There is thus formed at the side of the dibble a slight depression in which the left hand so buries the plant to the first leaf that the roots do not bend upwards but remain directed downwards; the dibble is then drawn out obliquely, the hole containing the plant is filled up with fine earth, and the space round it is somewhat pressed down with both hands so that the roots may not stand in a hollow. At the same time a slight depression is made round each plant in which the rain water collects.

In order that the plants may be set at regular intervals the labourers receive at first a piece of wood with which to measure, but after a time practice enables them to measure accurately by the eye alone, and the implement is then only used occasionally and to prevent mistakes. The more practised labourer always undertakes the first of the two rows standing close to each other, and after him the less skilful may accomplish the setting of the second because he has merely to take care that he hits the middle between two plants in the first row. If, in making the hole for planting, a stone or a clod of earth is felt below, this obstacle to the development of the root must be removed.

The setting of the plants is best commenced with a cloudy sky; in clear weather, the work should not be begun before three or four in the afternoon, because in the heat of the day the plants wither so much and then do not easily recover.

It is always better to set plants in a somewhat dry soil than in one too moist. If the soil is too wet the sides of the hole made by the dibble, especially in loamy ground, become too hard, so that the root cannot penetrate them, and such plants perish. The plants are thus always to be set in dry weather, but watered as much as is required, and if a shower then follows it is far more serviceable than if the setting had been delayed till after it.

Immediately after the setting, whether the ground be moist or dry, it must be watered, in order that the earth may adhere to the roots and that the plants may suffer no lack of nourishment, in a proper state of solution. If rainy weather succeed the setting of the plants, this one watering will suffice, but in dry weather the watering must be repeated until it is perceived that the plants have taken root in their new abode. This usually occurs within from 3 to 5 days, and it may be considered to have taken place when the centre leaf becomes stiff, of a dark green colour, and evidently larger.

The watering should be done only early in the morning, or better still in the afternoon after the heat of the day is over, and it is best managed with a small pot discharging the water from as short a distance as possible. It is bad to allow the water to stream forth in great gushes or from a great height, for in this way the earth when it dries becomes very hard and presses the stem together. The same thing takes place if the plants are watered under too hot a sun. When the soil is loamy it is well to sprinkle, the surface watered, with fine dry earth to prevent the formation of a crust round the plants on the drying of the mould. This should be done immediately aftering the watering.

Watering three or four times is generally sufficient even in dry weather. Too much watering is rather hurtful than beneficial; for the earth thus always becomes harder, especially when spring water must be used; many plants become sickly and die, and in the moist clods of earth worms are bred which injure the stems by feeding on them.

Worms of different sorts often work great havoc in the tobacco plantation. Many planters on this account set the plants very thick and when remonstrated with reply that the worms will soon thin them. If the worms would just eat away those plants that stand too near to each other, there would not be so much to object to in the practice, but this is not the case, and such plantations frequently exhibit spots completely bare whilst on others the remaining plants stand so thick as mutually to hinder development. Measured distances should therefore be observed, and the places of those plants which perish from drought, worms, or other injuries should be carefully supplied with fresh ones.

At the longest, eight days after setting, the rows must be all looked through and all vacancies supplied. At the first replanting, if it occurs within a short time, recourse may be had to the hotbed; choice must be made, however, of the most vigorous plants. But if one has a sufficient quantity of the early transplanted plants, it is always best to use them even for the first resetting. Later resettings, as has been already remarked, are only successful in their results when vigorous plants that have been early set are employed.

Against the injury done by worms, when they are already present, there is no real antidote except that of searching them out and destroying them. Manuring with powdered lime, as before mentioned, appears to be the most active treatment. When the tobacco is set in ground manured with fresh stable dung the worms breed most numerously, it is therefore best to put vegetable matter between it and the tobacco plants.

But not only are the roots and the stem of the tobacco injured by insects, the leaves also are frequently attacked. The springtail and the snail may be kept away by the same means as have been suggested in the case of the hotbeds. Much more dangerous are the green leaf worms which quickly pierce through many leaves, and must therefore be immediately sought out and destroyed.

Moles, which often undermine whole rows of plants, must be caught if possible.

Other animals which trample the plants or injure the leaves must be kept at a distance from the plantation.

The young plants may be killed by night frost, which congeals their sap and bursts the cell tissue. But if the leaves of the core have not suffered, which is sometimes the case when those already developed lie over them and protect them, the plants may yet recover even though attacked. The sprinkling of the frosted plant with cold water in the morning, immediately after the injury, has often been tried with success.

When after the transplanting there is reason to dread hoar frost, which usually only occurs through radiation, the best plan is to cover the little plants with other leaves. The same protection is also of essential service to the freshly set plant against the too strong heat of the sun. Large plantations far removed from dwelling houses may be best protected from injury by hoar frost by lighting towards morning smoky fires at distances of some hundred paces along the edge of the field whence the wind blows, and thus spreading a good deal of smoke over the field, which lessens the radiation towards the clear sky, and thus prevents the too great cooling of the field.

The precautions necessary to protect the tobacco plants from injury by strong winds have been already given in the description of the suitable situation of tobacco fields.

More dangerous than all the rest for the tobacco plantation is a hail shower.

If soon after the setting of the plants a shower of hail or very heavy rain occurs, which bruises the young sprouts and moreover plasters them over with mud, it will be necessary to wait for a time till the ground has dried somewhat.

After some days it will be apparent whether or not the plants can recover, and the number of those of which the leaves at the core are uninjured should be ascertained. If they predominate they must be freed with the fingers from the earth sticking to them, and the hardened earth around them must also be somewhat loosened where there are hopes of recovery. In such cases only the bruised plants require to be replaced. If, however, the number of dead plants predominate it will be best to harrow up the field and reset it afresh.

If the tobacco plantation be visited by hail at a time when the plants have already unfolded their stem leaves, and when the resetting of the field is impossible, though there is still time enough before the harvest for newly developed leaves to attain maturity, the loss, no doubt, is great, yet still a tolerably abundant return may be obtained by cutting off the stems close to the ground, when new shoots abounding in leaves will immediately be put forth, one of which, the most vigorous of course, must be left as a stem, the rest being removed, as otherwise the leaves would be very small.

If the plantation be visited with hail showers shortly before the harvest, the damage will be very great, for the torn and bruised leaves, even if gathered, will lose in weight and in value, and such a misfortune will press very heavily on every planter who has not insured his field against damage by hail.

TREATMENT OF THE TOBACCO PLANT TILL THE TIME OF HARVEST.

The tobacco plants require great attention during their growth, and this must be given if an abundant crop is to be obtained.

This attention consists in hoeing, heaping, nipping off the tops and the superfluous leaves.

THE HOEING.

This process is intended to loosen the soil round the growing plants, and to destroy weeds. It has an important influence on the promotion of growth, for by the loosening of the soil the entrance of air to the roots is facilitated, the penetration of the moist night vapours, and especially the capacity of the soil for absorbing moisture, are promoted; further, the spreading out of the roots is favoured, they themselves are more subjected to warmth and disposed to greater activity, and lastly, the weeds, which took up room, consumed the nourishment intended for the plants, and hindered the extension of their roots, or the beneficial influence of the light, are destroyed.

The human hand is the most perfect implement for loosening the soil round the stem of the growing plant, and for eradicating the weeds, whilst guarding against injury to roots or leaves. This style of labour however is too costly, and it is therefore generally replaced by the use of the hoe.

The tobacco plants are usually twice hoed; this first takes place about 14 days after transplantation, or as soon as they have struck root in the new soil and have unfolded some small leaves. It is best effected when the soil has first been moistened by rain, and then somewhat dried on the surface. The second hoeing takes place when the plants are rather more advanced, but before they shoot up to their height, and when the weeds begin to increase to a great extent.

Sometimes, when the surface of the earth is much hardened by violent rains followed by bright sunshine, so that a crust has been formed, the loosening must be repeated more than twice. The same is the case when the field is very much overrun with weeds, and these seem likely to get the better of the legitimate inhabitants of the soil.

Heaping.

This consists in heaping the earth round the stem of the plant, and is done when the latter has already begun to push up in height.

Heaping is accomplished with the hoe, or better still, by hand. First the ground is hoed, the weeds destroyed, and then the earth is cautiously raised round the plants, so that the leaves which have already attained some size may not fall over and be injured.

The object of heaping is to give to the plant a firmer hold, to promote the shooting out of the crown-roots, to protect the deeper roots more from drought, and thus make the plant less dependent on rain.

Nipping off the Tops and the superfluous Leaves.

In most of the native plants the stem quickly grows to its height, whilst the leaves of plants from Pennsylvania, Holland, Gundi, &c. first form a bunch close to the ground, because their stems only shoot up after the unfolding of the flower buds. The number of stem leaves on these remains limited to what the plant had developed at the appearance of the flower buds, and the growth of the leaves ceases to progress vigorously as soon as the plant devotes its principal energy to the formation of its organs of reproduction.

The tobacco plant endeavours with great vigour to fulfil the mission of its life, namely, to blossom and bear seed; but as our principal aim is the abundant production of leaves, we must intervene, and give to this activity a direction corresponding with out wishes.

If we wish to raise large vigorous leaves we prevent the plants from flowering by breaking off the buds or the blossoms, which obliges a portion of the sap taken up to be expended in the increase of the substance of the leaf. After the removal of the flower buds a slackening of vigour is observable at first, this however does not last long, and the leaves soon freshen up again, and the activity of the root is very much heightened in order that the plant may carry out, in another manner, the object of its life, which has been thus interrupted. The plant, robbed of its crown, pushes forth within a few days, on the shoulders of most of the leaves, buds which we call "suckers," which are destined to blossom and bear seed. Luxuriant plants are often found with these suckers even before the removal of the top buds. But these buds must be carefully and repeatedly removed, because they take up the food which, according to our endeavour, should serve for the development of the leaves, put the latter into shade, and narrow the space.

Many species of tobacco, as those of Pennsylvania, Ohio, Holland, &c., on the development of the flower buds, unfold so large a number of leaves that they cannot all receive sufficient nourishment and attain the growth desired. Early set and luxuriant plants can bring more leaves to maturity than late set and sickly ones. The latter must, therefore, be allowed to retain fewer leaves than the first. If heavy leaves intended for the manufacture of snuff be raised, their number must be still more restricted.

In accordance with the circumstance above mentioned, the bushy plant of Pennsylvania, Ohio, Holland, Gundi, &c. may retain from 16 to 22 leaves, the Hungarian from 10 to 15, and the original Galitzian from 8 to 12.

The condition of the plant and the object of its culture regulate the number of leaves to be removed with the flower bud.

The object aimed at in removing the flower buds, namely, the increased growth of the leaves, will only be fully attained when the exact time for its execution has not been neglected, and this time is when the upper leaves especially are still small, light green, not perfectly organized in their constituent parts, or, as the common phrase is, while they are still young.

In general the flower buds are broken off as soon as they begin to appear from amidst the "top" leaves which surround them. These leaves, which are still small, arrange themselves at this time round the stalk only, as the sun's influence is strong, and if they are held asunder with the fingers the flower bud within can be discerned. This is, as a rule, the most fitting time for nipping off the bud.

This nipping is done only during the strongest noonday heat, when the leaves, in consequence of the evaporation, are dry and languid, so that they cannot be injured by being touched, moved, and moderately bent. The operator approaches the plant as closely as possible, carefully spreads out somewhat with the fingers of the left hand the little top leaves which are to be left on the plant, and with the nails of the thumb and forefinger of the right hand nips off the still soft and juicy under part of the flower stalk by the exercise of moderate pressure. One must go to work with much caution so as not to tear or otherwise injure the still tender top leaves, for these injuries remain, and increase with the increasing growth of the leaf.

Along with the flower-buds, leaves will be removed in accordance with the before-mentioned peculiarities; from the Pennsylvanian, Dutch, &c., from 4 to 8 will be taken; from the Hungarian from 3 to 4, and from weakly plants a still greater number.

At the time when, according to these rules, the removal of buds and leaves is to be executed, the tobacco plant is still of low stature, the height of the stem being, according to the species of tobacco, from below a foot to a foot and a half. The height of the stem is not to be so much considered as the other signs, given above, with regard to the presence of the flower-buds. Whoever keeps in view the fact that the tobacco plant has attained its full number of stem leaves on the appearance of the flower-buds, and that it commonly bears more leaves than can be allowed to remain on it in order to their perfect development, will not regard the height of the stem, but will endeavour, at the proper time, to relieve the plant of whatever is superfluous, thus preserving, for the use of the leaves, the greater portion of its nourishment, instead of suffering it to be consumed in the production of a high stem and blossom. And, indeed, it is observable that very soon after the removal of the flower-buds, the bottom and middle leaves become of a lively dark-green colour, and increase in size as well as in thickness. But the effect of the operation shows itself strikingly on the upper leaves, which, if the removal of the blossoms did not take place at the right time, would always be smaller than those beneath, whilst they now rapidly increase in length and breadth, often attaining the same size as the centre stem leaves, but in thickness, richness, and substance greatly surpassing them. Thus well-formed top leaves have always relatively greater weight, and are selected either for twist or specially for snuffs of high price.

By nipping off the buds at the right time the stem is prevented from growing immoderately high, and at the utmost rarely exceeds from $2\frac{1}{2}$ feet to 3 feet, while, as the leaves stand close together, the wind can never cause so much damage as in those plantations where, in consequence of neglecting to remove the bud, or being too late in doing it, the stems have shot to a great height, the leaves are at a distance from each other, and mutually injure each other by fluttering about. Not unfrequently many tall plants are either uprooted during a storm or quite broken down.

The top leaves of plants thus allowed to grow tall always remain small, and the upper ones are of such inconsiderable size that it is considered a great mistake to mix them among the rest of the stem leaves, and yet they should not, as we have seen, come greatly behind the middle leaves in size.

The nipping of the buds is frequently performed too late, namely, when the blossom has already opened, and when the top leaves are no longer tender and of a light green colour, but have assumed a tough consistency and a dark green hue.

In this case the removal of the buds contributes little to the growth of the leaves, it still however promotes the absorption of mineral matter by the leaf-substance, and also aids in the ripening of the leaves.

People often go very unskilfully to work in removing the buds, taking hold of the plant by the top and tearing this off. The fragments of leaves at the upper end of the stem show the defects of this proceeding, a proceeding which has injurious consequences, for these leaves, of which only small shreds remain, would have afforded a considerable quantity of most valuable substance, had the suggestions offered above been attended to.

It is therefore obvious that due attention to the removal of the buds at the proper time is one of the most important conditions of good husbandry.

But, as has been already observed, this process is not available in the manner described for all species of tobacco, but only for those which have for their object the raising of large, vigorous, elastic leaves suited for the manufacture of snuff or of cigar covers, or of twist and ordinary cut tobacco. The timely removal of the buds is therefore to be specially attended to in the original Galitzian and Fünfkirchner, then in the Hungarian, Galitzian, Theiss, Debrecziner, Szegediner and Syuloker varieties, as also wherever the product is to be reckoned in the category of ordinary garden leaves.

The culture of garden leaves which are destined for the fabrication of the finer kinds of smoking tobacco, and where size and elasticity are not of so much consequence as the attainment of a bright yellow or pink colour and an agreeable aroma, is differently conducted.

The garden leaves become finer and better when the plant is permitted to fulfil all the functions which nature assigns to it, especially when it is not hindered from blossoming. The stems then attain to greater height, the leaves are at a greater distance from each other on the stem, the sun can get better at them, and can contribute to the elaboration of their sap, and there are no artificial obstructions to injure the flavour and the aroma.

The finest garden leaves are raised in Verpeléth and Debrö, because the plants are allowed to blossom and bear seed. The bright yellow colour is also only attained by attention to this mode of treatment. If the tops are removed before blooming, much larger leaves are obtained, but even in sandy soils their colour is always a darker brown, and in delicacy of substance as well as in flavour and aroma they are much inferior to those in which the plants are either not cut at all, or at least not till after the blossoming has taken place.

In order to raise garden leaves of certain districts, as for instance those of Csetnek, Vek O'Gyala, &c., and Transylvania, in a vigorous condition and of light red colour, the tops are removed, but not until after the plant has bloomed; then the flower is severed with a sharp knife.

The finest, early set, and most vigorous plants are to be selected for bearing seed; they are of course not cut, but careful attention must be bestowed on them; in particular, no "suckers" must be permitted to grow on them, and further, all the side spikes of the blossom must be removed, and the production of seed limited to the head, in order to obtain large, full grains with plenty of vitality in them.

The suckers, the appearance of which was formerly described, must be broken off as often as they are observed to increase. The first time will be when the buds are removed, when all the leaves of the plant are examined, and the buds or byshoots nipped off. According as the condition of the plant is more or less vigorous, this process will be repeated earlier or later till the ripening of the stem leaves, and will require to be performed twice or thrice.

These young juicy suckers are not in general gathered and dried, because they cannot be allowed to remain long enough to attain any size or maturity; they are therefore thrown on the ground and left there to wither, but on no account must they be allowed to rest on the leaves, where they stick, and might communicate rottenness to them.

Whilst removing the buds and superfluous leaves, attention should be directed to the sickly plants. Disease is shown by the leaves becoming rusty, wrinkled, or stunted.

The circumstances under which disease most frequently occurs are those which indicate that change of soil is desirable. Rust, however, also shows itself after wet, cold weather, or when the roots have been corroded by a lump of dung. If at the time of hoeing any still young plants exhibit a stunted growth, they are to be pulled up and their place supplied with vigorous pricked out ones. In larger plants the uppermost, tenderest leaves are commonly the first to turn rusty.

Such plants may often be saved from complete destruction by breaking off the stunted top, after which the roots strike afresh and keep the lower leaves in a condition to be available. Larger rusty leaves are gathered and dried at the time of the general pruning, as they can be used for ordinary cut tobacco.

Small white rust stains, which only appear here and there on the surface of the leaves, are not however considered injurious; usually such stains, which are called "spickel," on a sudden change of weather, attack even the best leaves—such as are highly prized as cigar covers.

Weakly yellowish plants, not advancing rapidly in development, can only be brought to greater vital energy by the removal of a large portion of the top.

If the plants selected for bearing seed should afterwards display stunted or rusty leaves, they must at once be pruned.

The operation referred to must never be undertaken when the tobacco is wet by rain or dew; we must, on the contrary, wait till the leaves are somewhat limp, in order to avoid injury.

Seed Plants, Harvesting, and Preservation of the Tobacco Seed.

In treating of the selection and rearing of good tobacco seed we have learnt to know the qualities requisite in the seed-plant, and all that is needful as regards the pruning or removal of superfluous leaves has also been stated. It is asked what is to be done with regard to the leaves generally, and when should they be removed? These questions are easily answered, if it is considered that the leaves are the plant's organs of breathing, and that they have the function of changing the juices absorbed and preparing them for further vital activity. So long as they are able to perform this office they should not be separated from the plant, because their activity is necessary to the development of the seed. But when they have attained perfect maturity their utility ceases, they bend down very much, and assume a yellowish colour; for as soon as the plant ceases, by means of respiration, to absorb the carbon from the atmosphere and to form chlorophylle, the green colour gradually changes, under the influence of the light of the sun, through disoxidation, into a yellow, passing over into what is called Xanthophylle.

The leaves of seed plants are therefore allowed to ripen somewhat more than those of the pruned plants; so soon, however, as they begin perceptibly to change the green colour for yellow, they must be gathered gradually in rows from the bottom upwards, till at last only the naked stem with the head remains standing.

By this time the plant has blossomed, and has only to perfect the seed and bring it to maturity. Often, the perfect ripening of the seed does not take place till several weeks after the removal of the top leaves, and if the plants are not early set the harvest frosts which usually take place in the month of September, seldom allow it to ripen completely. In plants of American origin, the development of which in all their phases is of slower progress, the earliest setting possible is imperative, in order to obtain both ripe leaves, and a larger amount of available seed.

When the outer husk of the seed capsules has become quite brown and dry, and the stalk is already withered up, the seed is ripe, and on opening the capsule will be found uniformly hard and of a brown colour.

If the capsules are then allowed to remain any longer on the plant they spring open and discharge the seed; the capsules are therefore gathered when the abovementioned signs appear, and when it is observed that some have already opened. Seed capsules, moistened by rain or dew, should not be gathered until they have become dry again.

All the seed capsules of a plant do not ripen simultaneously, therefore only the ripe ones should be taken, the rest being allowed to remain sometime behind.

The capsules when gathered should be spread out on an airy place to dry perfectly.

When it is observed that the greater part of the capsules are ready, it is a good plan to pluck off the inferior or imperfectly ripe ones, and set them aside, but to cut off the entire head with the good capsules remaining on it, and hang it up in a place suitable for drying it.

Sparrows are very fond of devouring the ripe seed, and therefore, both while it remains in the field and afterwards during the process of drying and preserving, their depredations must be guarded against.

After the lapse of some weeks the grains of seed become perfectly dry, and may be removed from the capsules; but they may also be kept there till the next sowing time if the heads have been cut off and hung up.

The shelled tobacco seeds may be kept in any suitable vessel which permits the passage of the air, but they must be protected from mice, and placed in a dry place.

The marks of good seed are, grains large, scentless, glossy on the surface, and moreover heavy; and on crushing the grains between the fingers the mass feels oily and shows the inside to contain white capsules.

The germinating powers of the seed is destroyed by age, overheating, moisture, and fermentation.

The tobacco seed retains its powers of germination during several years, yet seed from one to three years of age is to be preferred to older. Seed of a year old furnishes indeed the greatest number of plants, but these are but weak. When, for example, the seed, from drying up, becomes heated, or when it has in any way become moist and then again dried and remains lying, it loses its germinating power. The same result ensues through fermentation, when the seed is not sufficiently dry, and becomes heated while lying heaped up. If it smells it has either already fermented, or the process which produces mould is going on, and the germ is either dead or weakly. This circumstance usually occurs when the seed is kept in a damp ill-aired place. Where large quantities of tobacco seed have to be preserved it will be well to sift them over from time to time in an airy place.

GATHERING THE TOBACCO LEAVES.

The harvest consists in gathering the ripe leaves.

The leaves of a properly cultivated tobacco plant do not all become ripe at the same time; they attain this condition by degrees, and indeed in the order of the circles in which they are successively developed. Thus, the bottom leaves are first ripe, next the middle, and lastly the top leaves—called by the Dutch trade names, sandgut, erdgut, bestgut.

When all the leaves are gathered at the same time, as is still frequently the case, a portion of them must be either unripe or over-ripe, according as the harvest is early or late.

Leaves taken whilst still unripe become discoloured, thin, and powerless whilst drying, and as, in particular, they have not received a full measure of mineral substance they exhibit through repeated and violent fermentation a great disposition to decompose, and yield a tobacco of inferior quality, nay often quite worthless.

The over-ripe leaves also lose in value, because the substances absorbed with the descending vital sap gradually diminish in consequence of the exosmose, and hence the substance of the leaf becomes porous, loses in elasticity, shows greater aptitude to decomposition, and, at the least, assumes an unpleasant light colour.

Only tobacco taken at the proper stage of ripeness yields an article of commerce satisfactory in every respect.

The tobacco leaves are to be considered as ripe when their organic matter is perfectly elaborated, and is saturated as much as possible with inorganic matter, because this last is destined to preserve the first from decomposition and to contribute to the firmness of the tissue.

The different species of tobacco show their ripeness in different ways.

The sandgut exchanges its light green for a yellowish colour, and bends down very much towards the ground. It is gathered when yellow leaves appear here and there, and indeed all the leaves belonging to this class are taken away at the same time with the lumpsel, in order not to go too much about the plantation to the prejudice of the more valuable leaves yet remaining on the plants.

Thus some of the leaves of the sandgut, which besides only yields common cut tobacco, are allowed to become too ripe while waiting till the lumpsel, or the transition leaves between the bottom and middle leaves, has attained the necessary degree of maturity. This last, when ripe, is remarkably thick and stiff; for the rest, as in the sandgut, it is the brighter colour that indicates its fitness for being cut down. The sandgut is usually ready soon, after the removal of the flower-buds, and is thus already gathered when the bestgut begins to develop.

In about 8 to 14 days after the removal of the sandgut the erdgut is ripe. The colour, which at first was bright green, is now changed into a somewhat dull yellow, the leaf becomes observably thicker and tougher, in the daytime it droops, and in the cool of evening can no longer stand upright. In a good soil the change of the dark green colour to yellowish is not very striking, whilst on the luxuriantly grown erdgut leaves yellowish spots are observed, which give them a marbled appearance; they break readily in bending, if not much dried, and stick to the fingers.

At the time of the ripening of the erdgut leaves, especially in the evenings when the weather is still, a perfume like honey may be perceived in the plantation.

The erdgut is gathered when it exhibits those signs of ripeness already quoted, and all the leaves belonging to this class are gathered at once, even though each individual leaf may not strikingly display the tokens. Exception is however made of plants which have been late set, and are behind the rest in growth, as their leaves are in consequence entirely unripe, and require time to bring them to the requisite stage of development and maturity.

The removal of the sandgut and of the erdgut leaves contributes so much to the progress of the bestgut, that it is often ripe eight days after the latter. The bestgut leaves, at first, fold together in clusters in the cool of the evening, afterwards with increased development they remain stretched out or somewhat turned up; in the ripe state, they present a thick compact mass, become stiff, and easily break on bending; they also become sticky and adhere to the fingers. The dark green colour remains, or is changed into a somewhat brownish one, and the surface of the leaf is sprinkled over with brownish or yellowish spots like blisters, which sometimes degenerate into spots of rust. These stains give the leaf a strikingly marbled appearance, it moreover often becomes wrinkled on the surface and bent inwards and outwards. With bestgut over-ripeness is of less consequence, and it is therefore not cut down until the signs of ripeness are almost universal; then the whole is cut at once. Nevertheless a degree of ripeness so considerable that many of the leaves have become yellow, must be avoided.

In the culture of leaves for the manufacture of cigars, snuff, twist and ordinary cut tobacco, the plants of Hungarian or American origin show tokens of being ripe exactly in the order here described. Accordingly, in leaves of these sorts, sandgut, erdgut, and bestgut should be gathered at three separate times.

The original Galitzian plants are, in this respect, somewhat irregular, and indeed it is only the lower leaves of the sandgut which are earlier than the remainder; the rest ripen about the same time as the bestgut. These leaves are therefore usually gathered at two periods. As the stem leaves of the original Galitzian plants only bring a high price when they are heavy and substantial, care must be taken that they are rather over-ripe than unripe when cut, for being intended for snuff, if they are not sufficiently ripe they yield, as do all other leaves, a raw material unsuited for this purpose, and can only be used for ordinary cut tobacco, by which means a considerably smaller price is obtained for them.

As regards garden leaves, those which by drying are intended to attain a bright yellow colour must be allowed to ripen fully. The sand leaves present those tokens of ripeness which have been formerly described. The change of colour of the middle and upper stem leaves must not be limited to individual yellow spots, but the whole surface of the leaf, as well as the side ribs, must be bright greenish yellow like a ripe lemon. When not quite ripe, the garden leaves become, with good management, brown; with bad, they remain greenish, ill-coloured, or become quite black, but never attain the bright yellow colour desired. The ripeness of the stem leaves does not take place all at once, and therefore the harvest must be taken at three or four different times. In order that the larger and more vigorous garden leaves may attain a pink hue, a greater degree of ripeness must be aimed at than is required for the ordinary leaves; the signs of ripeness are the same as in those last described, only it will be necessary to wait till all the leaves of one sort are more uniformly yellowish. The gathering also must be made at three or four periods. Moreover, the garden leaves only attain their fine perfume when they have reached perfect ripeness.

Of course, early set plants ripen sooner than those set late, and on a plantation, where the setting lasts for a considerable time, the harvest of one kind of leaf can only extend to such time as the leaves take to attain perfect ripeness.

The removal of the leaves from the plant is effected by taking one leaf after another as near as possible to the stem, with the fingers of the right hand, grasping the midrib and breaking it off by pressure from below.

As where the leafstalk issues from the stem sideways at an angle, the vessels on each side form a joint, and as no very close connexion exists at this place, the arrival at maturity is shown by the weak coherence of these portions, which a very slight use of force will suffice to separate from each other. This slight adhesion of the leaves to the stem is therefore one of the tokens of perfect ripeness.

In many districts the planters have a bad custom of severing the leaves with a knife, so that larger or smaller portions of the wood of the stem remain on the leafstalk. These portions of wood are called tails. From what has been said, it is evident that there is no need of a knife to detach the leaves from the stem, and that far less is it desirable to sever any portion of the wood of the stem in the operation; for justifications of this bad practice, as, for example, that the tails assist fermentation, &c., will not stand examination, and are only to be considered as a cloak for the dishonest purpose of increasing the weight of the leaves by the portions of wood adhering to them.

The leaves after being separated are carefully laid down on the left hand, with the upper or sunny side lowermost, and as soon as a sufficient quantity of them has been collected, they are carried to the spot previously appointed, where they are arranged in small heaps, the upper side still downwards.

The severing of the leaves can only be effected when these are not wet either with rain or dew, and when they are somewhat dried up. The work should not therefore be commenced early in the morning, and in stormy weather it must be intermitted, as otherwise many leaves will inevitably be torn.

When the distances between the rows are not large enough to admit of a passage round, the luxuriant condition of the plant causes difficulties in getting the gradually ripening leaves out of the thicket. In such cases, the sand-leaves are frequently lost, because it is scarcely possible to pass between the plants without injuring the large yet unripe leaves. Under such circumstances, it is best to employ well-trained children to gather the sandgut, by creeping about right and left between the rows, and spreading the ripe leaves on the path.

Where, as in Hungary, the individual planters have very large sufaces of land to till, the requisite operations, as the frequent hoeing, the heaping, the topping and thinning of the leaves, and afterwards the gradual gathering of the ripe leaves, are seldom performed at the right time, but frequently some portions are still being hoed whilst in other spots the leaves are already quite ripe, all which occasions great diversity of labour and expenditure of time.

The consequence is, that in plantations where the proprietor is wanting in experience, or where the requisite number of hands for the various operations are not forthcoming, the work is not executed properly or in right time, and in particular the gradual gathering of the leaves according to the ripening of the different kinds is quite neglected. A usual accompaniment of all these circumstances, particularly when the setting has been too late, or the weather unfavourable so as to retard the plant's development, is the fear of frost, and then frequently the whole of the leaves of all sorts are gathered at once. Sometimes, in such cases, only the uppermost two, or at most four bestgut leaves, called the top leaves, are left for some time on the stem, whilst all the rest, namely, the sandgut, erdgut, and a portion of the bestgut, are gathered at once and become mixed together.

In this way it cannot be otherwise than that one portion of the leaves is either unripe or overripe, which must occasion loss one way or the other. Another mistake is from the mixture of the different kinds of leaf, as sandgut, erdgut, and bestgut, which, both in their special qualities and in their management during drying and fermentation, not less than in the mode of their employment, and hence in every respect, differ widely; and this senseless, yet frequent mixture of all the leaves gathered, must be considered as the grossest error of the native mode of culture, an error which can scarcely afterwards be repaired by the most careful and laborious sorting, as the different kinds when mixed present few marked characteristics of appearance whereby they may be separated, whilst individually, according to their nature and their degree of ripeness, they require different treatment.

If then one is not in circumstances to follow closely the rules laid down on this subject, and is obliged to delay the gathering of the sandgut till the period when the erdgut and the lower bestgut leaves are also sufficiently ripe, one may yet, by attention to the following suggestions, prevent the greatly injurious mixture of leaves, and this without requiring more labourers.

The examination of tobacco leaves can never be effected so quickly, narrowly, and at the same time so easily as at the time when they are still growing but just about to be gathered. It is then easiest to determine which leaves, on account of their size, beauty, and freedom from blemish, are adapted to the most remunerative use of cigar covers (super-excellent), and, on the other hand, which, from being small, insignificant, torn, or rusty, can lay no claim to high value, and therefore rather belong to the refuse.

The most judicious labourer, or the planter himself, should therefore go foremost between the rows, picking out only the best uninjured leaves which he considers of first class quality, and breaking them from the stem one by one without meddling with the still unripe uppermost leaves, those of inferior quality, and the sandgut. After him should come his children or assistant workmen and take the sandgut and the refuse which he has left behind, and the gathered leaves being separately spread, the principal sorting is completed in this extremely simple manner and without any increase of labour.

A further important advantage of this process is that the attention of the planter can by means of it be specially devoted to the best leaves, the best portions of the drying-houses can be assigned to them, and a better system of drying applied to them, whilst the inferior quality of leaves, which are only available for the manufacture of cut tobacco, may, so long at least as places for drying are limited and new methods of the process not commonly exercised, be treated according to the manner hitherto in use.

After-growth and Suckers.

At the gathering of the sandgut and erdgut the suckers are taken away at the same time, in order to promote the nourishment and development of the bestgut leaves as much as possible. If the setting of the plants was early effected and the weather tolerably favourable, all the stem leaves have attained the necessary ripeness by the middle of September at the farthest, and there then remains sufficient time for the development of the after-growth, especially when a good late autumn without frost favours vegetation. The roots of the pruned plants already far spread resume their activity after the ripening of the leaves, and furnish the stem uninterruptedly with an abundant supply of nutritive sap, in order to develop new buds and to attain the aim of the plant's existence.

The stems of the plants are therefore soon again furnished with suckers. At the harvest of bestgut two or three of the most vigorous shoots are left, but the rest are removed, in order to bring the others to better development. But if the bestgut is ready to be cut by the end of the month of August or the beginning of September, it is more expedient, immediately after the removal of the bestgut, to cut down the stem quite close to the ground, because then more vigorous shoots will appear, the leaves of which sometimes attain considerable size. In order however to attain this end, the smaller shoots which will soon produce flower buds must be pruned down, and the number of leaves too must be restricted, only from three to five leaves per shoot being allowed to remain.

The larger leaves of this shoot are called after-growths, and if sufficiently ripened and successfully dried and manipulated, they may attain to average quality and be at least available for use.

If however the smaller shoots are not lopped off or their tops removed, or if the time be too far advanced, the leaves remain but small, and would scarcely reward the trouble of gathering them separately and further attending to them. In this case when the leaves are ripe the entire shoot, the flower having been first plucked off, is taken with the small leaves hanging from it and in this state is subjected to drying. But when this last process is completed the leaves must be severed from the stem the woody part of which is of no other use than to be burnt. This product, the lowest in value, is called " geiz."

The ripeness of the after-growth may be discovered by the same signs as that of the bestgut, though from the advance in the season the proper degree of maturity can rarely be waited for, and as drying seldom succeeds well late in autumn, these leaves of the second harvest usually attain but an inferior quality.

THE DRYING OF THE TOBACCO LEAVES.

The drying of the leaves consists of a series of operations every one of which has for its aim a special object. These operations are:—

1. *The wilting of the leaves.* The green leaves are so stiff, and at the same time their surfaces are so sticky, that they do not bear handling until they are wilted or withered.
2. *The stringing of the leaves.*—The wilted leaves are then fastened together by means of strings or of sticks. Arranged in this manner they are subjected to all the next operations, and are exposed to less mechanical injuries.
3. For a certain time the leaves strung together on strings or sticks are left in *close suspension*. The end in view in this operation is the change of colour, consequent on a chemical decomposition of the vegetable matter. The leaves are therefore left in close suspension, to prevent too rapid an evaporation of water, because then the leaf would keep its green colour on drying. The yellow colour obtained by the *close suspension* is not to be confounded with the bleaching of the leaves when exposed to the direct action of the sun. This must be prevented as much as possible, as the leaves or parts of leaves which change their colour and dry up by the direct action of the sun get very brittle and powerless. This constitutes the so called sun-blight.
4. *The suspension in the sun* is applied after the change of colour has taken place, and its principal object is to dry and shrivel up the fleshy midribs, which may else in the subsequent operations originate mouldiness (called core-blight).
5. Drying of the leaves. It is only after these preliminary operations have been carried out that the leaves are to be deprived of their moisture; and their drying must be effected in a gentle manner. The operation is completed when the midrib has become perfectly dry and stiff.
6. After this the dry leaves are hung up closely together, and left so for some time, in order to equalize and ripen the whole produce.

Drying Houses for Tobacco.

The best and most beautiful leaves, when the drying is completed, must often, in sorting, be treated as refuse, because, during the necessary manipulations or in hanging up in the open air, they have got torn, or because they have become affected by core blight or are otherwise discoloured. And thus the proportion of leaves of best quality very frequently melts away to an inconsiderable quantity, because it is only occasionally that individual leaves are under such favourable conditions in the process of drying as not to be seriously injured.

In the usual mode of drying, the weather plays a very important part, and on its favour or the reverse hangs the result.

It is the condition of the thick juicy midrib in the centre of the leaf and covered over by it which more particularly constrains the planter to expose the closely strung leaves for an excessive length of time to the open air and the sun; if this midrib could be overlooked, the drying of the leaves might be attainable without the assistance of sun and wind, and would thus escape the injuries associated with them.

In order then to render the process of drying as much as possible independent of the weather, and to regulate it by our own will, it is necessary in stringing to give such a position to the midribs that they shall not be covered up by the leaves so that the air may have free passage to them and may effect their rapid drying.

Were the midribs, as is practised in Holland, slit up with the knife, and the leaves then fixed on sticks, the drying of the ribs would proceed rapidly, and without danger even in a narrow space, and the necessity of exposing the leaves too long in the open air would at the same time be avoided.

We have then to examine two different modes of tobacco drying, both approved of, the one employing strings the other sticks for the suspension of the leaves.

In both cases the preparation of the leaves chiefly takes place in the drying house, and it is therefore desirable to become acquainted with the essential conditions of a drying house. These are :—

1. It must afford complete protection against rain and wind, and hence also, when necessary, must admit of being closely shut on all sides, but it must not be damp and mouldy;

2. It must be provided with openings which can be closed, so that ventilation may be maintained at pleasure in the directions required;

3. The drying house should keep out the external heat, and be lighted sufficiently to allow of the execution of the necessary manipulations in the interior by the light of day;

4. There must be suitable arrangements for the suspension of the tobacco leaves; lastly,

5. The quantity of tobacco leaves to be prepared in it must be proportionate to its size.

All buildings or parts of buildings which fulfil these conditions are suited for the purposes of tobacco drying.

Once that the process of drying has begun, the leaves must never be allowed to get wet, for if they only get sprinkled with a few drops of water and the moisture cannot at once be expelled, it attacks the substance of the leaf, and immediately blackish stains appear which are never effaced.

If the moistening be greater, and lasts longer, the leaf at once rots.

If draughts have entrance into the interior, the leaves flutter about and are injured. In such cases, as also when the drying house is too hot, the drying proceeds too rapidly, from which cause the quality is much impaired.

Thus, if the substance of the leaf dry too quickly, the leaves shrivel up and become hard, the slow and equal decomposition of matter which must precede the attainment of a fine colour is interrupted, the substances dry up, and it is difficult, by the subsequent attraction of moisture, again to excite them in such a manner as to bring on a uniform change. In consequence, the leaves which have dried and hardened too quickly, always remain greenish or streaked, and even on fermentation, only assume a dull fawn colour, whereas for leaves used in covering cigars a light brown or bright reddish and uniform tint is required.

In order to attain the quality desired, the leaves, in drying, must never be so far deprived of their moisture that they rustle or feel hard, still less must they retain or attract a superabundance of it. From the treatment prescribed they should always remain soft and pliable, so that they can easily be unfolded and stretched.

As a rule therefore the drying house in dry, hot, windy weather, as well as on rainy days, should remain shut. On the other hand, ventilation is sometimes necessary, especially with fresh tobacco, or after damp weather, and therefore both in the walls of the drying house, and also in the roof, suitable places for openings must be fixed upon; and where the object is to admit access, gates and doors must be contrived; where light is in question, windows.

Drying houses, with walls of masonry or stamped clay, require more apertures than those wainscotted with boards or wickerwork: and in the case first mentioned, especially, the ventilators must be near the ground, because the lowermost space is usually damp, and the tobacco hung up in it is apt to mould and rot.

A thatched roof is the most suitable for a tobacco drying-house, because it excludes the external heat.

At convenient places in the roof, windows or skylights should be made, so that a cross draught may be obtained by their transverse position.

It is a good plan to place doors in both gables near the roof, in order to be able, if required, to produce a draught along the whole length of the building.

In order to get the full use of the space afforded by the building, it is necessary to place frames at distances determined by the length of the strings. These frames consist of perpendicular strong poles, the lower ends of which are sunk in the ground and the upper fastened to the rafters. They are the props of the horizontally-placed laths, which are hung with strings of tobacco. The horizontal laths are fixed to the upright poles with strong nails, or clasps of iron or wood; sometimes, also, the perpendicular poles are pierced at measured distances, and in the holes thus made, strong wooden pegs are inserted, the horizontal laths being made fast to these by plaited willows or strong cords. At the top, the horizontal laths are fixed to the timbers of the building. The distance of the horizontal laths from one another is regulated by the length of the tobacco leaves. When these are large two feet are usually allowed; when small, one foot. The lowermost lath must preserve a greater distance from the ground on account of the bending of the lines and the injurious moisture of the ground.

The distance between the poles of the frame is regulated by the length of the lines or rods used for stringing the leaves.

The most suitable length for the strings is nine feet, and from this length a judicious planter should not vary, in order to prevent the manipulations from being too difficult, and in particular, because, as we shall hereafter see, with longer strings the final suspension cannot be suitably executed, and, lastly, because in the transition from drying on strings to drying on rods, the existing fittings may remain, provided an additional frame between each two already there be added. Lastly, as regards the size of the drying house; if the whole of the crop is to be dried and prepared in it from first to last, this size should be very considerable. On an average, seven lbs. of small leaves, and above ten lbs. of the larger, may be dried in each cubic yard of space in the drying house.

The early-cut erdgut will soon dry, and may be at the close of the process or in final suspension, when the later ripe bestgut is only just gathered. In this case the drying space may be partially occupied twice, and, according to circumstances, even three times, with fresh tobacco.

As in Hungary and Galitzia there would be a difficulty in erecting such large drying houses as would be required for drying an entire crop on the plan here proposed, it may be arranged that only the best uninjured leaves, those adapted for covering cigars, &c., should be dried in this manner, whilst sandgut, refuse, and after-growths, which are usually only fit for use as ordinary cut-tobacco, may still be prepared in the manner practised heretofore; that is to say, may bn suspended on the sun frame in the lobbies of houses, under the eaves of roofs, on hedges, walls, &c.

Assuming that the drying house is 18 feet broad, the height of the side wall 6 feet, and the height of the roof equal to half the breadth of the building, then for each acre under cultivation we shall require from 14 to 18 feet length of drying houses, in order to be able to house conveniently the most valuable leaves of an average harvest.

Plate I. Fig. 5. represents a drying house in the usual architecture of the country, with stamped clay walls, half of it arranged for being hung with strings, the other half with rods. Plate II. Fig. 2. is a ground plan; Fig. 3. a transverse, and in Fig. 1. a vertical section is shown.

Structures of a different kind, such as are used in Holland, are shown in Plate II., Figs. 4–11.

In the drying house, the necessary space must be allowed for the different operations. In buildings more than 24 feet broad, a passage of from 5 to 6 feet wide is made in the centre throughout the whole length, leaving at the side walls only so much space as to allow free access to the ventilators, doors, and windows. If the building is narrow, a passage 3 to 4 feet wide at one of the side walls runs through the entire length.

When a new building is erected for tobacco drying, attention should be given to the arrangement of the interior, so that joists, crossbeams, rafters, &c. may not be placed in such a manner as to come in the way of the frames, but, on the contrary, that they should be opposite to them, so that advantage can be taken of their position.

Empty stables, lofts, sheds, barns may be arranged for drying tobacco, provided the foregoing requirements are observed. In case of necessity, a booth may be very quickly erected on the ground on a spot not lying low, in which the tobacco may be dried without risk. A section of a structure of this sort is represented in Fig. 7, Plate I.

Small planters of narrow means may make a shift by placing timber obliquely against the walls of their dwelling houses, lathing these and covering them with straw; thus prolonging, as it were, the roof of the dwelling house down to the ground, and arranging the sheltered space thus obtained for tobacco drying, in the manner depicted in Fig. 6, Plate I.

It is thus obvious that no very costly erections are required to obtain the space requisite for tobacco-drying, and it would be indeed deplorable if henceforth, from indolence, valuable tobacco leaves were exposed to the scorching of the sun, whilst at the same time lofts, sheds, &c., perfectly suitable for the purpose, stand empty, although they could be prepared and used for the approved mode of drying with small labour and inconsiderable expense.

As the foregoing pages have thus set forth the essential conditions of a judicious drying of tobacco, we shall now occupy ourselves more closely with the individual manipulations.

The Wilting of the Leaves.

Before the actual drying of the leaves can take place, there are still some preparations requisite; these consist in tying up and fixing the leaves on cords or rods.

After the leaves are removed from the plant, the first process which they have to undergo is that of withering or wilting.

In treating of the gathering of the leaves it was mentioned that on being broken off from the plant they were spread for a time in the spaces allotted for paths, in small orderly piles, with the upper or sunny side downwards. Their withering in these heaps on the field proceeds most rapidly, and at the same time most successfully, when the state of the weather so far favours exhalation that no water drops are formed on the surface of the leaves in the interior of the heaps. Sunshine and moderate breezes are the chief agencies in affording these favourable conditions. In order, however, that the leaves on the top may not be scorched by the continued influence of the sun, the small heaps which are arranged in rows must be covered with something light, for which purpose the spurious shoots, of which there is never any lack, may be best employed.

The sandgut and erdgut, especially those of the garden leaves, must be brought under cover the same day on which they are gathered, because the moisture of the dew would exert an injurious influence on them. The bestgut of heavier sorts of leaf may in dry weather remain overnight in the fields in heaps, but if rain at all threatens, these also should be brought under cover.

The green tobacco leaves are very easily injured by pressure, and wherever they have suffered from this there originate dark green or black stains which cannot be obliterated.

The leaves must, therefore, never be handled except by taking hold of the midrib, and in placing them, moving them, &c. the greatest care must be observed. Accordingly the bringing home of the leaves from the field by means of waggons can scarcely be accomplished without injury even with the greatest amount of caution, and how much more is this the case when the matter is carelessly performed, as, for instance, when the drivers sit on the green leaves and urge on their horses. If, then, the distance of the plantation from the drying houses be not too great, the best plan is to remove, at least the most valuable portion of the crop, by handbarrows.

With heavier leaves the necessary degree of withering can rarely be attained on the field, and therefore the leaves which have been brought home, if they are still fresh and stiff, must be further allowed to wither before they can be hung up. For this purpose they are arranged on the ground in rows of about 6 inches in height, the leaf-stalk downwards, and the point upwards, and with the sunny side of the leaf also lowermost.

In these little heaps the leaves are allowed to remain lying until the evaporation of the sap is so far advanced that the midribs have become somewhat soft and flexible. For the withering of the leaves an airy locality is best suited, so that an open shed, a barn, or the drying house itself, the ventilators being opened to allow of the easy escape of moisture, will answer the purpose. But if these precautions are not observed, or the heaps are of too large a size, the fresh leaves laid down to wither become heated, and the moisture which cannot escape rapidly enough condenses into drops which make the leaves in the inside of the heap wet on the surface. This is called the sweating of the tobacco leaves.

This sweating is always injurious, for, partly through heat, and partly through the dissolving power of water, a rapid decomposition of the surface of the leaf commences, the individual substances lose their water of hydration, combinations, separations, and even crystallizations take place, in which they become more and more rich in carbon, because, in the formation of the escaping carbonic acid, the organic substances give forth a smaller relative quantity of this gas than of oxygen and hydrogen, which are employed partly in the formation of carbonic acid, partly in that of water. This process goes on in the moist porous mass of fresh tobacco leaves very quickly on the admission of air, and may be perceived by heat, formation of water, and the separation of carbon.

A green leaf hung up by itself, in a dry place, loses the water contained in it by evaporation merely, and remains, in the dry condition, green. But when the chemical process just described takes place, in the moist heap of leaves, the colour of these changes, in consequence of the relatively greater presence of carbon, which increases with the progressing decomposition of the leaf, and the original green colour becomes gradually yellow-green, yellow, brown, and at last black.

Moist albumen very easily passes into a state of decomposition. Over-ripe leaves, which, in consequence of exosmose, have lost a portion of their mineral ingredients, change colour more rapidly, and immediately decay when laid in heaps to wither; but even properly ripened leaves rot if they remain too long lying in the heaps.

It is thus obvious that by withering the tobacco leaves in heaps their chemical and physical properties will be essentially modified if they become heated and sweat, and that if a certain point be overstepped the leaves will be entirely spoiled.

The leaves which in withering sweat and become warm may indeed easily attain a yellow or brown hue, but they always become stiff and brittle after the drying, and this to a greater or less extent according as they have gone to a greater or less extent through the process above described.

Such tobacco leaves as are to be manufactured into cigar covers, or twist, and require flexibility, or those which, on account of their richness, are suited for making snuffs, must not be exposed to sweating.

Sandgut, which, usually, are somewhat over-ripe, bear sweating very badly, because they readily putrify. The same is the case with fine garden leaves grown in sandy soil, they cannot bear sweating, and if subjected to it, immediately become black.

Sandgut generally, and fine garden leaves in particular, where the soil is light, are only allowed to wither a little in the field, and as soon afterwards as possible are hung up.

With regard to garden leaves of stronger soil, a moderate amount of sweating does no harm, because in this way a bright red colour is more readily attained. These leaves are allowed to lie in the heap till they have attained a yellowish colour. Attention, however, must be given to the matter, for the riper leaves turn yellow sooner than the less ripe, and in warm weather the process goes on more rapidly than in cold. Thus when the heaps become somewhat warm they must be turned over, and the leaves that have already become yellow must be taken out. These manipulations occupy, according to the circumstances described, from two to eight days.

With heavy middle and upper stem leaves it is, as has already been said, only a suitable amount of wilting that is desired, without heating and the formation of water; the heaps must therefore be turned over from time to time in order to prevent these injurious appearances, and to prevent evaporation. According to the degree of ripeness attained, the withering in the field, and the state of the weather, the midribs become in two or three days so far flexible that the next operations can be proceeded with.

The Stringing of the Leaves.

This operation is differently performed according to the use of strings or of sticks.
Two different methods are employed for fastening the leaves on strings.

1st method.—For the fixing of tobacco leaves a needle is employed, from 10 to 20 inches long and about ¾ths of an inch broad, with sharp point and blunt edge. Plate I. Fig. 1.

With this needle the midribs of the leaves are pierced at about an inch from the thick end, and then pushed close together. When the needle is almost full of leaves these are pulled on to the cord threaded through the eye, and this process is continued until the requisite number of leaves have been strung.

Most of the tobacco planters make the strings too long, and fill them too full of leaves. The manipulation with long strings is always difficult and induces many deteriorations of the leaf. A further disadvantage arises from the great bending in the strings when too long, causing the leaves to slide towards the centre and to stand too close there. On strings too long the wind exerts a strong influence, swinging them violently about so as to stain the fresh leaves with black stripes or spots where friction takes place, and to tear those that are in a drier condition. Other disadvantages of the strings being too long will hereafter be adduced. If the strings are nine feet in length, all the manipulations may be performed without difficulty. When the material is good and well manufactured they can be tightly stretched so as not to give way in the injurious manner just described even in a stiff breeze. In order however to save the erection of so many supports, strings 2, 3, nay sometimes $3\frac{1}{2}$ fathoms in length, are frequently employed, partly from custom and partly from not perceiving the disadvantages entailed by the practice; but if these disadvantages were duly taken into account, the conviction would speedily be arrived at that the saving of some props, in this manner, is dearly purchased by the deterioration of the produce.

When the leaves are hung too thickly their surfaces come into immediate contact at the midrib, and the deteriorating process already described in the sweating of tobacco goes on. The leaves change in the vicinity of the midrib both chemically and physically, become yellow, brown, or black in colour, lose their firmness of texture, become brittle after drying, or if the moisture is long in effecting its escape both surface and ribs decay. This appearance is called blight, core-blight, or rib-blight (brand, herzbrand, rippenbrand); it constantly attends a too thick hanging of the strings, and is very frequently observed in Hungarian leaves as well as in Galitzian. The leaves are especially liable to this injury when they are arranged, as is customary, close to each other and always with the same side in one direction. The leaves lie as we have seen with the sun side next the ground. The labourer takes a parcel of leaves in his hand, places them now with the sun side upwards, lifts them leaf by leaf in his left hand, and pierces them with the needle, which the right hand directs. Thus leaf after leaf comes towards the labourer with the sun side next him, and so too on the needle and the string. But in drying, the leaves have a tendency to curl up in this same direction, so that frequently several leaves get rolled up together, and the strong moist ribs become enveloped by the surface of leaves also moist, which, as will easily be perceived from what has already been said, speedily causes decay, if the leaves are not disentangled from each other. Every planter has, no doubt, undergone this sad experience more than once, yet from habit, and from the desire of sparing lines, the practice of too thick hanging is continued.

This last disadvantage is effectively prevented if the leaves, in hanging up, are constantly arranged so as alternately to present to each other opposite sides. After a little practice the piercing of the leaf from the rib side is as quickly effected as from the other, except that a good deal of time is consumed in turning every second leaf, which must make the process slower than the common mode. This however may be rectified by the labourer's having two heaps at hand, with the leaves of each turned in opposite directions, when he has only to bear in mind never to take a leaf from the same heap twice consecutively. The stringing will then proceed in the manner desired without any trouble in turning the leaf.

2nd method.—It has already been remarked that in this process the object to be accomplished is to turn out the portion of the leaf which is most difficult to dry, namely, the thick, juicy, midrib, so that it is not covered by the rest. The two sides of the leaf, when withered, are disposed to bend and fold over to the sun side, leaving the midrib so plainly exposed that it can be pierced sideways with the small needle (Fig. 4. Plate I.) in the manner represented. It is only to be observed that the leaves, first in the field and afterwards in the drying house, under the cautions laid down, must be well withered, so that the sides, when the leaf is grasped in one hand by the midrib and raised from the heap, may bend together of themselves or with a little help from the finger. If the leaves, however, are stiff and not well wilted they do not allow themselves to be folded together with such ease, and the side ribs break when it is attempted to bend them.

The stringing should not therefore be proceeded with until the leaves are so far wilted that the side ribs allow themselves to be bent with facility. In the stringing, it must always be observed, that the midribs be turned to one side and the surfaces folded together to the other.

One principal condition of a good preparation of the leaves consists in not stringing them too close, but always leaving them at a distance of from half an inch to an inch apart, because otherwise the yet moist surfaces press on each other, and from too close and long contact pass into injurious decomposition and rottenness; the leaves are therefore strung on the string till it is two thirds of its length full, when it is hung up in a light place for adjusting; there, the leaves throughout its entire length are removed to exactly equal distances, and if any of them have their sides unequally folded they are put straight, particularly if they fold round and cover the midrib.

Fig. 2. Plate I. represents a line, strung, and adjusted in the manner here described.

Another way of fastening the leaves together is by means of sticks or rods. This is to be preferred to the use of strings, as the leaves are less subjected to mechanical injury.

The rods used in drying tobacco leaves must be at least half an inch in thickness, but if two inches in diameter, they will be quite suitable for large leaves. They must not be very crooked, and they must be smooth on the surface, and free from knots. They are most easily managed at the length of 5 feet, and as 3 inches of this, on each side, has to be reckoned for resting, the distance of the framework comes to be 4 feet 6 inches, exactly the half of what was mentioned as most suitable for strings.

The drying rods are pointed at the thick end.

The hazel-nut bush offers the straightest, smoothest speile, not requiring the bark to be removed, and possessing great durability. Other species of wood of which the bark is rough must be prepared with the knife, so as to offer a smooth surface.

Young acacia shoots, not much bent and of sufficient length, are likewise well adapted for tobacco drying. Not less useful speile are supplied by the stronger willow wands.

In Lower Hungary and the Banat, where there is a want of all these kinds of wood, the stems of the sunflower offer an excellent substitute for the purpose in question. The sunflower is cultivated to some extent as an oil plant, and its stem, which hitherto has only been used as fuel, would soon be found of essential service in the drying of tobacco. This plant grows to the height of more than 6 feet, and is generally straight to the height of 5 feet; near the ground it attains a thickness of from 1 to above 2 inches, and as, besides, its capacity for sustaining weight has been tested and found more than sufficient for the purpose, it may be employed with advantage in the drying of tobacco. For this end, after the seed is ripe, the stem of the sunflower must be cut down close to the ground, then exposed to drying, and afterwards preserved in bundles of 25 pieces, in sheds under cover. At the most seasonable time, which is winter, when the planters have little else to do, the stems are prepared as drying rods. First, the thicker end is pointed with a sharp knife, and then the stick is placed on a fixed measure of exactly 5 feet in length, and the portion which exceeds this is cut off. Lastly, the projections where the leaves grew must be smoothed away with the knife. The prepared rods are then stored in the drying house till required. In the same manner rods of wood are prepared, only in this case, instead of a common knife, stronger implements must be employed.

Wooden rods must also be perfectly dried before being used, for the moisture of the rod would result in the rotting of the leaf hung on it.

In order to prepare in the improved manner the beautiful uninjured leaves selected for covering cigars and as first quality, at least 2,000 pieces of drying rods are needed for each acre of surface in tillage.

The wooden rods may last for several years, and those formed of the stems of sunflowers if used with care will remain serviceable for more than one year; besides, these last grow every year, and by suitable cultivation may constantly be produced in any required quantity.

In Holland, not a single tobacco leaf is dried on a string; the use of the rod is universal, because the manipulations there employed in drying can only be perfectly executed with rods; and these moreover have the advantage of requiring no hooks, while the expensive provision of lines is entirely avoided.

The mode of proceeding is in general the same as in drying on lines, the same principles and cautions are to be observed, except that a few manipulations have to be altered in order to meet the conditions involved in the use of rods. The whole manipulation consists in the slitting and the stringing. For the slitting of the leaves when sufficiently withered, each labourer requires a knife made for the purpose, which consists of a fixed blade, sharp, and coming to a narrow point, $1\frac{3}{4}$ inches long, by $\frac{1}{2}$ inch broad, with a handle 2 inches in length.

The labourer seats himself on a low stool, and places on his lap a sufficient quantity of tobacco leaves to allow him to move his hands easily, putting them with the sunside downwards, and the strong ribs or nerves directed towards the left hand. About ¾ of an inch from the point, he grasps the knife between the forefinger and thumb of his right hand, so that both these fingers hold the blade at equal distances, and thereby project somewhat over the edge, whilst the end of the handle rests on the palm. The left hand takes the uppermost of the leaves spread on the lap by the end of the midrib and raises it up somewhat, whilst the right hand, about an inch from the end, sticks the strong rib with the knife held by the same fingers, so that the finger points embrace the rib somewhat on both sides, but the other fingers which are rolled together lie on the leaf, so that the hand does not shake. The left hand is now raised somewhat drawing the leaf upwards, whereon the closed fingers of the right hand slip down the midrib, and a slit is made through its centre, without the right hand having been itself drawn away.

The fingers of the right hand must constantly remain lying on the midrib, whilst the upward motion is executed by the left, because otherwise the knife starts aside. The same happens if the attempt be made to cut the slit with the right hand alone.

By raising the leaf with the left hand, every injury to the leaves remaining in the lap is avoided.

This care and manipulation must be observed at first till the labourers are expert in slitting the leaves, afterwards they find no difficulty even with another position of the finger, and in a few days attain such a proficiency that an industrious workman can slit in an hour more than a thousand leaves.

The slit is only made long enough to string the leaf easily on the rod, and it is thus from 4 to 6 inches in length. If the leaf be split further up, the two sides separate too much from each other, and these far slit leaves are in consequence, difficult to work without injury, when the midrib has to be removed in manufacture.

The slit leaf is laid aside in the following manner:—When the slit has been made of the required length, according to the foregoing directions, the midrib is grasped somewhat more firmly by the right hand, and without further assistance laid on the right side on a heap, in which it is to be constantly observed that the leaves must be placed upon each other as before. Whilst this goes on, the left hand grasps the uppermost leaf on the lap, and the manipulation proceeds as described.

At the slitting of the leaves their sorting should also be accomplished. In this operation the beautiful, the large, the small, the more or less ripe, the torn or otherwise injured, are separated from each other, the labourer placing the leaves to the right side, yet according to their varying qualities in different heaps, which occasions no loss of time, as he has each leaf separately before him, and can look at it.

The slit tobacco leaves are taken, when sorted, by other labourers, by whom they are put on the rods.

The labourer intrusted with this operation seats himself on a low stool, and takes a suitable quantity of slit leaves on his lap, so that the points come to lie to the right hand. Then, on the left hand, a rod is leant in an oblique direction, so that the pointed thicker end remains about 1½ feet above the labourer's leg, whilst the thinner blunt end rests on the ground. A simple stand to steady the rod may also be prepared to prevent any delay in the work, although it is obvious that the seated labourer without much exertion can easily reach to the point of the rod with his right hand. On this properly placed rod the leaves are so strung that they alternately turn in a different direction; if the first has the sunside outermost, the next has the under one.

The reason of this mode of proceeding has been already more minutely explained in treating of tobacco dried on strings.

For the purpose of this alternation of the leaves, the labourer has to turn each second leaf in the fingers of the left hand, which is done at the moment when the leaf is raised from the lap in order to be transferred to the fingers of the right hand, whilst these are introduced into the opening slit. When he has thus got four, six, or eight leaves (even numbers are to be observed) in the fingers on the right hand he sticks them at once on the rod which is leant against his leg or otherwise steadied.

In order to dispense with the turning of each second leaf the labourer takes two heaps of slit leaves of the same sort, but turned in opposite directions, the one with the sunside uppermost, the other lowermost, and places them on a convenient stand before him, and taking the leaves alternately from these heaps, he arranges them as before mentioned on the fingers of the right hand, and then puts them on the stick in small bunches at a time. The strong leaves remain at first at the thick end of the rod closely pushed together, and as many are to be put on one rod as shall thickly fill about a third of it, whilst the longer thinner portion is allowed for the present to remain empty. On a rod 5 feet long, with sandgut and small leaves generally, about 50 may be strung; with erdgut and bestgut from 30 to 40.

Close Suspension of the Leaves.

The tobacco strings when adjusted are immediately suspended in the drying house, and indeed in the lower portion of the frame, close together, so that the leaves on adjoining strings touch, but do not press on each other. The green leaves thus hung bear without injury for several days this close proximity without becoming heated and sweating; the process of change, which has been more particularly described in treating of the fading of the leaves, goes on in this way slowly and regularly, and we have it in our power to regulate their tint at pleasure.

The dark green colour of the leaves soon changes during the close suspension into a faint green, then always becomes lighter, fades into yellow, yellow spots always increasing on the inner surface, and if the leaves remain long in the close suspension, they become bright yellow, brown, and at last black. It is evident that this appearance depends on a slow decomposition of the leaf substance, by which the carbon always comes more to light.

If the decomposition of the substance oversteps a certain degree, so that the material of the fibre is also affected, the dried leaf is deprived of its elasticity, and may even, while still in close suspension pass into decay.

The close suspension must therefore be carefully watched, and the leaves allowed to remain in it only as long as is necessary, in order that, when perfectly dry, the light brown or pink hue required for leaves that are to cover cigars may be attained.

This is, in regular drying, attained when the leaves in close suspension have so far changed the dark green colour that they begin to pass into citron yellow, or as it is usually expressed, to break out into yellow.

The leaves must remain in close suspension till they become yellow-green, or are speckled with yellow, which may take a longer or shorter time, according to the natural condition of ripeness, the kind of tobacco, and the state of the weather. Yellow leaves are not put at all in close suspension. Fully ripe rich leaves begin to colour earlier than unripe and poor ones. Warm moist weather favours the chemical process to which the close suspension conduces, so that by this means the desired change of colour is earlier attained.

According to the circumstances just alluded to the tobacco leaves in close suspension will require from 5 to 14 days, in order to attain the requisite change of colour as uniformly as possible.

At the part of the drying house where the close suspension is performed, the ventilators must be kept open, in order by continual change of air to favour the evaporation. A strong motion of the green leaves by the wind is however faulty, because, whenever they press or rub against each other, black spots or stripes appear.

As soon as one parcel of leaves in close suspension has attained the change of colour desired, the strings must be removed further from each other, in order to stop the process of decomposition at the right time.

The same operation performed with sticks.—The strung rods, placed according to their various qualities, are immediately put in close suspension, being so arranged in the lower part of the frame, that the closely strung leaves at their thicker end hang to the right the first time, to the left the next, and so near that the edges of the first row come into close contact with those of the third, and those of the second with those of the fourth, and so on, without however exercising a strong pressure. This arrangement is necessary in order to save room.

The green leaves remain in close suspension until, as already described in the manipulation with the line, their colour so far changes as to burst into yellow.

Suspension of the Leaves in the Sun.

It is desirable that at this time calm, sunny weather should prevail, so that the strings prepared for drying by close suspension may be hung on the sun frame (Fig 1. Plate I.)

Should the weather however be stormy or rainy, the strings are merely to be removed farther from each other, so that by opening the ventilators the evaporation may be maintained as much as possible.

Only after the change of colour has been suitably prepared are the leaves in a condition to free themselves easily from superfluous moisture so as to dry well and securely; this drying succeeds best when the tobacco line is transferred immediately or soon from close suspension to the sun-frame.

On the sun-frame the strings must be hung sufficiently near to each other to afford some degree of mutual protection from the direct influence of the sun, especially on the edges of the leaves, and from the too rapid drying up and scorching of the leaves generally. In front and behind are hung strings full of refuse leaves, in order that the good tobacco may be screened as much as possible from the injurious influence of sun and wind. When the sun shines on the frame, the exhalation of watery sap proceeds so rapidly that the leaves often lose in one day more than the fourth part of their original weight, which object is much more slowly reached in the drying house. If good, sound, finely coloured tobacco is to be produced, the sun-frame must not be passed over.

The tobacco leaves remain on the sun-frame only until the ribs become soft and the remaining portion perfectly limp, so that each leaf hangs by itself without being hindered by the one next it from folding up.

If settled weather prevails and there are no high winds, it is very conducive to the after drying of the leaves to allow them to remain on the sun-frame for evaporation 2, 3, or even 4 days; but the strings must be placed closer from time to time, because the leaves always droop more as the evaporation goes on, so that they lose their close proximity, and lay bare their edges to the injurious influence of the sun.

It will be observed that the hanging in the sun is chiefly designed to draw forth the superabundant moisture of the midribs; for the rest of the leaf all that is wanted is that it shall be completely limp.

When the leaves are rather closely hung on the sun-frame they cannot so readily suffer injury from the wind if, as has been already mentioned, the first and last strings are filled with refuse leaves, and by shortening and lengthening the strings are hung higher and lower than the good leaves, while in all cases mats of straw and rushes are placed at the sides. It is found that thus even the points of the leaves remain quiet in a breeze; for when they flutter about they soon turn black, and a hot wind parches them too much and makes them hard. The necessary means of protection against wind must therefore not be lost sight of, but if it become too violent the leaves should immediately be transferred to the drying house.

The wetting of tobacco leaves by rain is under all circumstances injurious. Therefore when there is a threatening of this the leaves must be at once transferred to the drying house. If they have got wet already, while it is evident that the rain will last but a short time, it is best to leave them on the sun-frame till the shower ceases, but then the lines must be immediately removed to a greater distance from each other, in order that the air may penetrate to every part and volatilize the moisture. In continued rainy weather the lines which have got wet must be hung up in open sheds or near the doors and ventilators, in order that they may speedily part with the injurious moisture.

Dew is only hurtful to the tender garden leaves; those suited for cigars and snuffs may in calm, steady weather be left without detriment on the sun-frame overnight. But when the leaves are transferred from the sun-frame to the drying house, this should not take place till the moisture of the dew has been perfectly volatilized by sun and wind.

Long continuance on the sun-frame is very disadvantageous to the tobacco product. The edges of the leaves are burnt and become of a bright yellow. The highly prized portions of the surface dry too rapidly as far as the air penetrates, and remain of a green colour, while farther in, where the moisture cannot readily escape, the surface of the leaf, as we have already seen in regard to its sweating, becomes more or less decomposed, and black or brown in colour. Tobacco long exposed on the sun-frame always exhibits this inequality of hue in a striking degree, and it can never be wholly remedied by after treatment. The upper portions of the leaf, which have become hard and brittle through over-drought, are easily damaged by the slightest touch, or if the line be set in motion by the wind. Yet frequently, on the occurrence of a gale or a shower, these lines require to be hurriedly removed under shelter, by which means considerable injuries are unavoidable. And thus, even when the leaves have grown well and have been successfully housed, the mode of treatment usual in Lower Hungary and the Banat occasions the profit on perfectly uninjured and uniformly coloured leaves to be, at last, very small.

The same operation with sticks.—This process is executed in the same manner as the preceding. The rods strung with leaves are transferred to a sunny spot in the open air, placed on the frame erected there and moved so close together that the leaves of the respective rods touch.

In other respects, the same course is to be observed as in the use of the strings.

The carrying out and in proceeds rapidly, as each labourer takes several rods in his hands, while the placing of them in the frame is more readily effected than is possible with the use of strings.

The Drying of the Leaves.

When the object of suspension in the sun is attained, the leaves are brought into the drying house, where the process of their thorough preparation is to be completed.

Before the suspension however they must be loosened with the hand in case they have anywhere stuck or folded round one another, and be equally divided on the string, so that henceforth each may hang by itself and not become entangled with the one next to it. In order to secure good results this regulation of the leaves must not be neglected, and it should be confided to none but trustworthy hands and be closely superintended, for when the leaves remain closely huddled together, or when several have got entwined together, the enclosed moisture cannot escape, as the drying under cover proceeds but slowly, and the consequence will be nothing less than the decomposition and corruption of the leaf substance. If however the foregoing cautions be observed there is nothing to fear, and the preparation of the leaves will answer expectation in every respect.

In the suspension of the leaves in the drying house it must be further observed that those on adjoining strings must no longer come in contact. The strings are therefore hung at such a distance apart that an empty space of from 1 to 2 inches is left between the leaves, the hooks being placed from 6 to 12 inches from each other according to the breadth of the leaves.

In this operation of hanging the leaves in the drying house it is best to proceed from the top of the frame downwards, and to fill one frame completely before beginning on another.

In order to be able to hang up the strings at the greatest height required, easily, regularly, without much loss of time, and without risk of injury, each drying house should be provided with at least two double ladders, on which the labourers who undertake the suspension as well as those who hand to them the adjusted lines may ascend to the necessary height.

The tobacco leaves remain suspended in this manner until they attain the suitable degree of dryness, especially until their midribs become hard and woody.

At each different stage of the operation it must be observed that dissimilar leaves, as sandgut, lumpsel, erdgut, and bestgut, those moreover of different species, must always be kept apart from each other; for the sandgut dries rapidly, erdgut more slowly, and bestgut more slowly still, and the mixture of the leaves would also essentially increase the difficulty of the subsequent sorting.

The leaves must not dry rapidly, they should never nestle, nor feel hard to the touch, but always remain pliant and elastic, in order at last to attain a good quality and a fine colour. In hot windy weather therefore the drying house must remain closed, and if it is observed that the leaves have still become somewhat hard in the daytime, some windows or airholes must be kept open over night to attract dew for remedying the defect. The injurious effect of too rapid drying on the uniform colouring of the leaf substance has already been explained.

If, however, from continued rainy weather or thick fog, the leaves attract too much moisture, and if this lasts long, there is danger of their assuming too dark a hue, becoming mouldy, and finally turning soft, and even passing into decay.

Therefore, each time that rainy weather has been of long continuance, all doors, gates, windows, and airholes must be opened at its close, in order to free the leaves from the superabundant moisture. Those of which the substance has been affected by damp, or through a fault in the adjusting of the line, must, when the weather becomes good, be removed to the sun-frame, and a check be put to their being spoilt by promoting drought, while the position of those that had been rolled together or too closely pressed must be rectified. If however such faults as the foregoing be avoided, there is very seldom in our climate any danger to be apprehended from the injurious influence of external moisture; more frequently, indeed almost always, it is the too rapid drying which is to be guarded against in the manner described. A slight appearance of mould is soon lost on the admission of dry air, or it dries up without causing any damage.

The tobacco leaves are to be considered as sufficiently dry when, as before mentioned, their midribs are hard and woody, and when the rest of the surface has attained that pliancy and elasticity which fits it for being made up into bunches.

According to the state of the weather and the species of tobacco this process of suspension lasts from 3 to 6 weeks.

The same operation, but with rods.—Before the leaves are subjected to this process they are distributed over the entire length of the rod at equal distances from each other. The distances from rib to rib should in large leaves amount to perhaps an inch, in small ones to half an inch, so that the too close contact of the sides of the leaves may be prevented, and that, as the process of drying proceeds, each leaf may have room to fold together by itself. When it is observed that one rod is too crowded with leaves, some of these should be transferred to an empty one kept in readiness for the purpose, and from this are supplied such rods as at first have too few.

The rods adjusted in this manner are placed in the frame at such a distance from each other that the edges of the leaves no longer touch.

The hanging up of the rods can be managed from one stand-point to a pretty considerable height, as the labourer employs a wooden pole forked at the end in raising first one end and then the other of the rods to those laths which are above the reach of his hands alone. The same forked pole is employed in taking down rods from a high perch. Where the height is too great for easy work, even with the fork, the labourer stands on a double ladder, or on a strong board, placed at a convenient part of the frame across the horizontal laths.

A single labourer can always work independently with the rod, whilst for stretching and hanging up the lines two individuals are constantly required, and the bending of the lines always occasions difficulties and delay in the work.

For the rest, the same conditions are to be observed as have already been described in reference to the manipulation with lines.

FINAL SUSPENSION.

When the tobacco leaves have attained sufficient dryness, when their midribs are sapless and hard, but the sides of the softness, elasticity, and pliancy desired, precautions must be taken for their remaining in this good condition and not drying up further, which would be prejudicial.

This object is attained by the final suspension.

For this purpose the best sheltered frames are chosen, somewhat removed from the entrance door, so that draught can be easily prevented.

Originally the distance between the horizontal laths in the frame was measured by the length of the leaves, so that during the process of drying, those hanging uppermost, might not reach too far down upon those in the lower row. In the process of close final suspension, however, it is necessary to bring the horizontal laths so near to each other that the leaves in each row overlap each other about half their length. Thus, in the final suspension, between each two laths in the frame, according to the length of the leaves, one or even two more must be fastened in.

In arranging for the final suspension, the work is begun at the back side wall and successively carried forward towards the centre of the drying house.

It is well beforehand to stuff the space between the back wall and the frame with straw to the roof, to prevent injurious influences.

In the first instance the tobacco strings next the wall of straw are hung from beneath upwards, as shown in Fig. 8. Plate I., so that the points of the leaves stand inwards, while towards the outside the leaves form a wall or screen.

But each succeeding row is hung from above downwards, Fig. 9. Plate I., so that the strings are stretched as tight as possible and pressed against each other, in order to attain the requisite compactness of the mass of leaves. Fig. 10. Plate I., shows a completed arrangement of leaves for final suspension.

If the strings are longer than nine feet they cannot be so well stretched, and in the centre a bulging out takes place, which is a fatal obstacle to great compactness. The first and last rows, as well as those lowermost, should be formed of refuse, in order to afford the greatest possible protection to the valuable leaves in the interior. It is well to strew some straw beneath the frame, and also to protect the sides of it with thick straw mats, against injurious influences of either dry or damp weather, as also of frost.

Leaves which have become too dry and somewhat hard should not be subjected to this final process of drying till, by the use of rain water or by exposure to dew or mist, they have attracted sufficient moisture to be soft and pliant enough to be made up in bunches. If, however, leaves that are too damp, or such as have still pulpy midribs, are thus suspended, they will most probably pass into decay.

Thus, before being hung up, each string must be closely examined; if on this only a few unfit leaves are found with ill-dried midribs, they are plucked out so as not to be obliged to subject the whole line to further drying; the leaves so gathered are immediately arranged on a fresh line and treated as required by their condition.

In the frames for final suspension, constructed in the manner described and protected against injurious influences, the leaves lose nothing of their interior contents, remain elastic and pliant, the inequalities in the degrees of their dryness disappear, the colour becomes very fine; thus the preparation succeeds in every respect almost to perfection, and at the proper time, whatever the weather, the making up of the hands may be commenced.

The tobacco leaves therefore remain in final suspension until the time arrives for making them into hands and consigning them to the Government authorities.

The longer this process of final suspension lasts the more perfect becomes the product.

Erdgut leaves of plants set in the month of May will, under tolerably favourable circumstances, if treated in the manner laid down, be ready about the middle of October, and even earlier, for this final process of suspension.

The same operation, but with rods.—As soon as the tobacco leaves have attained the necessary degree of dryness the rods must be closely filled throughout the entire length, in order to proceed to the final suspension.

For this purpose, to the dry leaves on one rod pushed together towards its thin end, those of three or four others are added, a number of leaves being pressed firmly together by the tops of the ribs, so that the slits are not widened, and this is continued till the rod is quite full.

By this arrangement space and rods for tobacco newly brought in are at disposal.

The final suspension is prepared and executed as has already been described in reference to the manipulation with lines; it must, however, be apparent to every one that this important operation can be far more easily and perfectly executed with the rods than with the lines, because the laying up and pressing of the rods only requires one person, and because no injurious bendings and bulgings out occur, which in the case of hanging on lines do harm to the compactness of the mass of leaves.

THE SORTING AND MAKING INTO HANDS OF THE TOBACCO LEAVES.

This last operation in the preparation of tobacco, whilst in the hand of the planter, is of special importance, as regards the qualities and value of the produce.

Even when in the harvesting, and the stringing of the leaves, the most careful sorting possible has been employed, it scarcely ever happens that the leaves hanging on one string, or on one stick, are so equal that the same quality can be assigned to them all. Moreover there are always, with the best treatment and amongst the best parcels, some refuse leaves, for now and then from oversight, a torn, unripe, or over-ripe leaf will get mixed up with the rest, or it may afterwards by chance have sustained injury.

If this be the case with judicious management, how much more so with the usual mode of proceeding. The advantage of the rational treatment consists in this, that the proportion of refuse is limited to the utmost possible minimum, whilst by the usual mode of treatment it depends principally on influences beyond the reach of man to modify. If the result of the drying turns out well or the reverse, if the amount of refuse be large or small, in general it is not satisfactory.

In the trade, far more serious consequences are involved in the sorting of the leaves than can be satisfied by the mere separation of the refuse from the good. That is to say, it is desirable that the leaves should be duly separated according to their suitability to the individual branches of manufacture, nay even according to their suitability for finer or more ordinary manufactures, partly in order to employ each individual quality to the best advantage, partly in order to accomplish as economical a delivery as possible: and therefore the accurately sorted ware is always much better remunerated than the mixed, for as the manufacturer does not undertake the sorting leaf by leaf, the good ones placed with the bad, cannot be taken into account.

The aim of the cultivator must be specially directed to the production of a handsome and strong covering leaf for cigars.

The peculiarity of the cigar covers, in so far as this depends on the choice of seed, are:—

(a.) Suitable size. The covering leaf must not be too large, but if too small it is of no use.

(b.) A fine elongated form, so far indeed that the length of the leaf is almost double its breadth. It is considered the most desirable average size when, in the green state, the leaf for cigar covers measures about 26 inches in length, and 12 to 14 inches in breadth, and when the outline of the leaf approaches the form of an oval rather than when it ends in a long sharp point.

(c.) A surface as smooth as possible. Folded, bulging, or baggy leaves will not lie flat out when stretched, and are less suitable for cigar covers than the leaves whose ribs and cell tissue are regularly disposed in a flat surface; lastly,

(d.) The position and character of the ribs is also of special importance. If the side ribs are too thick, or run too much in points from the midrib, or if they are curved in an undulating manner, the cigars so covered have an ugly appearance, and such leaves are consequently less prized in cigar manufacture.

In very large leaves the side ribs are commonly too coarse at their commencement for the purpose, so that only the borders can be employed in the production of the finer cigars, on which account the middle sized fine ribbed leaves always have the preference in the trade.

The side ribs therefore should be as fine as possible, and not less than an inch apart; further, it is desirable that they issue from the midrib at right angles, and, without making any bendings, run to the edge of the leaf in a gently curved line.

The secondary nerves of the side ribs, if they stand too close together, or are too thick, may prove an obstacle in the production of cigars. This defect is, however, common only with ordinary leaves, when their side ribs are already of considerable thickness.

But the leaves of the very same plant exhibit very striking variety in regard to size, form, and quality.

As has been already mentioned, the development of the leaves proceeds gradually from beneath upwards.

The lowermost leaves are formed first, and as the organs of plants are feeble in their early stage, these leaves remain small and weak. Their shape is always more rounded off than that of the other stem leaves.

The more however the roots of the plant spread out and strike deep, the stronger is the nourishment of the plant, and the larger become the leaves, and indeed this enlargement advances until the plant begins to develop its organs of reproduction, and to devote its activity principally to these.

Accordingly, from the middle of the stem upwards to the crown, there is a decrease in the size of the stem leaves till they gradually pass into the tongue-shaped calyx of the blossom.

The centre stem leaves are therefore as a rule the largest, finest, and tenderest, and are distinguished by a corresponding elegance of form.

The uppermost stem leaves whose growth, as will be afterwards shown, may be essentially promoted by the removal of the flower buds, are always thicker and have coarser ribs, and a narrower and more pointed form.

The stem leaves are thus divided into lower, middle, and upper leaves.

In trade, the Dutch names for these are most frequently adopted, and the lowermost are called "sandgut," the middle "erdgut," and the upper "best gut."

The uppermost more developed leaves of the sandgut class forming the transition to the erdgut, and which, on account of their fineness and the thinness of their ribs, are sometimes highly prized as small covering or swathing leaves, are usually known under the name of "lumpsel." The transition leaves from erdgut to bestgut are distinguished by the name "zweifler."

"Geiz" (suckers) is the name given to the commonly small leaves gathered from the offshoots of the plant. Larger leaves of the second crop, of which more hereafter, are called (after-growth). On Plate III. are depicted several forms of leaves at one sixth the natural size, in the green state, yet mature.

Fig. 1. sandgut, Fig. 2. lumpsel, Fig. 3. erdgut, Fig. 4. bestgut, are all taken from one kind of tobacco, which is raised abundantly both in Hungary and Galitzia, and furnishes excellent cigar covers.

Less suited for the purpose indicated are the extremely large leaves represented by Fig. 6., with thick side ribs, commonly curved, sometimes nearer, sometimes farther apart from each other. In the process of drying, the side ribs do indeed invariably become crooked in some degree from the contraction of the organs, but in the species under consideration the defect is specially conspicuous, the ribs being so thick.

Such leaves are therefore usually employed as cut tobacco, or, if elastic, as twist, but rarely possess that richness required for the manufacture of snuff.

The species, the centre stem leaves of which are represented in Figs. 7. and 8., are very numerous in Hungary, and they have been brought thence to Galitzia. In Hungary they are called "fodoros" or curled American tobacco, being descended from a Virginian species. The leaves are baggy and puckered, have thick side ribs, sometimes standing very close upon one another, and running very obliquely from the midrib, thus chiefly affording cut tobacco only, and at rare intervals peculiar snuffs. It would be well if this species of the plant were entirely rejected.

On the other hand, a very good form of leaf brought from Cuba is represented in an erdgut, Fig. 2. The side ribs, which issue from the midrib nearly at right angles and relatively at wide distances from each other, become thin and tender in the dry state; the leaf, when the soil is powerful, attains a considerable size, is elastic, substantial, and in the manufacture takes on a fine dark brown colour. The leaves of this plant develop themselves quickly and come to maturity with corresponding quickness, they have therefore, especially when raised on a sandy soil which does not permit standing water, a pleasant smell, and burn well, and in the fabrication of cigars can always be employed with advantage either for the inside or the covering according to the size required.

Plants of Pennsylvanian origin appear very often with very narrow leaves, after the fashion depicted on the erdgut leaf, Fig. 15. These narrow leaves abounding in ribs have no profitable employment, and therefore their continued cultivation must be noted as undesirable.

Other sorts, resembling the Pennsylvanian, are obtained from good Dutch and Gundi seed. The erdgut, Fig. 17., is usually smooth or only moderately folded, and has thin well arranged side ribs; with good management it feels tender and silky, and accordingly is much prized as a cover for cigars when its extent in breadth is not too small, which, however, is of frequent occurrence in the cultivation of Dutch seed.

The commonest shapes of garden leaves are shown in Figs. 21. and 22., but if foreign seeds are cultivated these become modified. The size depends on the quality of the soil, and on the circumstance whether the plants are allowed to blow or not. In loose sandy soil, and when the plants come to flower, the leaves always remain small; in chalky, humose, loamy soils, however, the garden leaves attain, as in Transylvania, a considerable size, especially when for this end the blossoms are removed in time.

Whether at all, and in how far it is desirable to remove the blossoms of garden leaves will be hereafter more minutely discussed.

Figures 23. and 24. exhibit the forms of the cserbal leaves, called in Hungary kapadohany, in Galitzia bakun. These leaves have always a beardless stem, seldom attain any considerable size, are wrinkled and puckered, and encumbered with thick, undulating side ribs, and they ought to be rooted out completely.

First of all, then, the tobacco leaves must be separated according to their sort, sandgut, lumpsel, erdgut, and bestgut.

Sandgut only furnishes ordinary tobacco for cutting; if it has already been separated at the drying, it need not be sorted any more; it is therefore stripped from the lines or rods and bound at once into bunches, without undergoing any smoothing out of individual leaves.

The larger lumpsel leaves become, with good management, fine and elastic, they then yield a useful roll for cigars, and should therefore be separated from the smaller ones which are only employed as cut tobacco.

The most usual and the best leaves for covering are furnished by erdgut rationally dried. Faulty, discoloured leaves must be carefully sorted out, because they only find their use as spun or cut tobacco.

Bestgut may furnish smaller cigar covers, heavy material for the manufacture of snuff, spun and cut tobacco, accordingly it may be fine or coarse, beautifully tinged or discoloured. Rich and poor leaves should not be mixed together.

The foregoing, though but an incomplete sketch of the varied employment of tobacco leaves, should convince the planter that the demands of the manufacture are very numerous, and that the sorting must correspond to these as far as possible, in order to establish the value of the commodity.

For the attainment in the manufacture of the aim already mentioned, and in order to induce tobacco planters to use honest and accurate modes of parcelling, the general orders for classification of the Royal Tobacco Administration were fixed, according to which the acceptance and payment of the produce is determined.

As these regulations are of high importance to most tobacco planters, and as according to them the accurate sorting of the kinds of tobacco raised here is characterized with all precision, it will be well to present them here in their exact terms.

In judging of the sorts of tobacco to be paid for by Government, and of the division of these into separate classes according to their quality, the rule will be:—

1. Among ordinary leaves shall be included those suited for the production of common smoking tobacco and snuff. Of these belong:

To the class of excellent, those beautiful gold and uniformly coloured, perfectly ripe, uninjured, smoothed stem leaves which from want of elasticity, delicacy, or from the coarseness of the ribs or deficiency in combustibility, promise no abundant profit as cigar covers, but are exceedingly well fitted for the interior part of cigars, twist tobacco, spun tobacco, or heavy snuffs, if they are quite purely parcelled and absolutely free from after growths and top leaves.

To the first class of ordinary leaves belong all sound, ripe, uninjured, smoothed stem leaves fitted for good cut tobacco, or for the insides of cigars, or for common snuffs.

To the second class of ordinary leaves belong the smaller uninjured stem leaves; then the larger smoothed stem leaves, which have sustained some injury from wind or hail; the unsmoothed, uninjured, purely parcelled, dry, finer sandgut; lastly, uninjured but scarcely ripened stem leaves and ripe top leaves.

To the third class of ordinary leaves belong parcelled and smoothed stem leaves much decayed by hail, or greatly torn, or such as are discoloured or speckled with white, or of which the stem has somewhat decayed, though the leaves have been dried; further, small sized but available top leaves and sandgut; lastly, such ripe stem leaves as are streaked by the frost, provided that they are not black, but available for ordinary smoking tobacco.

Suckers may be offered unsmoothed, but must contain neither stumps (storren) nor wood, and must be ripe and fit for manufacturing purposes.

Fragments, torn and loose leaves, must be delivered pure and without stumps; impure fragments will be reckoned at half price, or still lower, according to the expense of cleaning which is to be afforded by the reduction; it is taken for granted that the material is perceptibly durable, material not possessing this quality, or not worth cleaning, is destroyed.

3. As covering leaves for cigars, only such are fit as have been dried on a judicious plan, are of sufficient tenacity, delicacy, tenderness of rib, of a pure equal colour, perfectly ripe and easy of combustion. The bunches which contain injured or otherwise inferior leaves will only be paid for as middling, nay, according to circumstances, as ordinary leaves.

In the first class of covering leaves for cigars rank only the perfectly uninjured, finest tender ribbed stem leaves of considerable and equal size, superior elasticity and combustibility, of beautifully uniform colour, affording a return of at least 50 per cent of fine covers.

In the second class are ranked those similar to the first in colour, delicacy, combustibility and fineness of ribs, but which only attain middle size, and promise a return of at least 40 per cent of fine covers.

In the third class are ranked the leaves which in fine and ordinary covers offer a return of at least 35 per cent, if the remaining portion be perfectly available as material for the inside of cigars. Muscatell leaves, which are available as covering leaves, will be ranked and paid for in one of the three classes according to the proportion of anticipated return in coverings, supposing them to be of such size that at least with half a leaf a large cigar can be covered.

4. Under fine and moderately fine garden leaves are comprehended those which are distinguished by perfume, delicacy, and their bright yellow or light yellowish brown colour, and are suited for the manufacture of fine smoking tobacco.

In the class of excellent, of fine and moderately fine, garden leaves, are ranked only specially distinguished finest, clearly parcelled, perfectly free from blemishes, fully ripe, uninjured, uniform stem leaves of light yellow colour, in Siebenbürger flame coloured.

In the first class of fine and moderately fine garden leaves are ranked the fine yellow, or light yellowish brown, clean parcelled, uniformly coloured, perfectly free from blemish, fully ripe, and uninjured stem leaves.

In the second class are ranked the less uniformly reddish or light brown coloured, fine, perfectly ripe, clean parcelled, uninjured stem leaves, then the slightly torn stem leaves possessing all the other qualities of the first class, and the top leaves and superior sandgut of fine bright colour.

To the third class belong all sound, parcelled, smoothed, common stem leaves, suited for the manufacture of ordinary smoking tobacco, then those much damaged by wind and hail, so far as they are still available for good cut tobacco.

General rules for all kinds of tobacco; ripe aftergrowths which attain the size and quality of stem leaves shall be ranked in the same class as these, according to the foregoing regulations, only they are not eligible as cigar leaves, excellent or first class.

In drying, the tobacco leaves roll and fold themselves together somewhat; it is however desirable that they should be flattened out as smooth and as much in their natural form as possible before being made up into bunches. The tobacco leaves stript from the lines or rods are therefore carefully unfolded one by one with the fingers and spread flat on a board. In smoothing out the leaves they are laid one above another with the sun side downwards, so that those already finished form a soft support for those which come after, by which means the manipulation succeeds better. The sorting must be simultaneously executed, and in this operation many things are to be considered.

The endeavour of the planter must therefore be principally directed to the attainment of the greatest possible uniformity in size, form, colour, fineness, position, and quality of the ribs, elasticity, and in other superiorities or defects of the leaves to be bound together in one bunch.

It is evident that if the leaves are taken at once from the plant and mixed up, when strung a great variety of sorts must appear, for the distinct separation of which, there is seldom sufficient space in the planter's chamber. In such cases accurate sorting is really a difficult task, scarcely an attainable one, on which account the bunches are then not so particularly selected, and very often leaves of all qualities are mingled, as finest cigar covers of erdgut with coarse ribbed, core-blighted bestgut; further, with sun scorched, torn leaves, even to inadmissible sandgut.

If however at the gathering, and afterwards at the stringing and drying, the rational plan has been observed, the leaves present themselves at this state already separated into the great divisions, as lumpsel, erdgut and bestgut, some of which have only a few subdivisions; as for instance, in erdgut, cigar covers first, second, and third class, superior and refuse; or in best gut, cigar covers, three classes; heavy superior leaves for snuff, lighter superior for other purposes, and lastly refuse. Even the refuse should be divided into superior and inferior, in order to reap the utmost profit possible from the higher price of the ordinary class of leaves.

Of larger leaves at most 50, of sandgut up to 100, should be tied up in each hand.

With the better class of leaves it is usual, when these have been already smoothed out and well arranged, to lift about half the quantity designed for each hand, turn them round, and replace them on the remaining portion in such a manner that the strong ribs on one side correspond in position to the leaf points on the other. This is done in order that from the middle of the bunch the prettier sunside may be turned outwards and that the ware may gain in appearance. Above all care must be taken to make the hands firm, and for this end, as well as the above-mentioned turning of half the leaves, attention must be paid to the due arrangement of the ends of the ribs with the leaf points, and then to the firm tying up of the bunch.

The hands must be fastened an inch from the ends of the strong ribs, and this must be done with twisted tobacco leaves. For the fastening of the finer garden leaves moistened maize straw is used.

Any one who does not well understand how to use the tobacco bands will never succeed in making up handsome bunches with the required firmness, and the leaves when further handled are apt to fall out. The following method should therefore be observed:—

The fastenings of tobacco are prepared from elastic after-growth and smaller refuse leaves. The workman lays two leaves on a table before him so that their surfaces somewhat overlap each other and that their midribs are turned in opposite directions; both leaves are then rolled together with the fingers. Thereafter the man grasps the one end of the roll between his knees and turns the other with his hands till the fastening has attained the requisite firmness and the edges of the leaves are completely turned in. In order to prevent untwisting, both ends of the string are at last taken together in one hand, whilst with the other the bending is twisted somewhat. These strings must not be thicker than half an inch at the utmost, and perhaps 12 inches in length; they should always be prepared in the evening for the day following, only they must not be exposed during the night either to draught or frost.

The leaves to be bound in one hand are, after this preparation, placed by the labourer, who is seated, in such a manner that the points hang down perpendicularly between his legs and the midribs project somewhat beyond his knees with which he holds them fast. A string is passed with both hands at the distance of an inch from the thick end of the midribs towards the body and by pressure of the thumb is stretched tight on the upper part of the bunch. Then the end of the string, held in the right hand, is formed into a loop, which is immediately fastened to the end at the left hand, and wrapped round as far as it reaches, when the fastening is made secure with a flat pointed piece of wood.

After this the ribs are somewhat pressed together and the hand is ready.

Of similar importance to the strict sorting of the leaves for individual hands is the accurate placing of the whole parcels of leaves as much as possible according to the similarity of the bunches, and in this matter the same things are to be observed as have been stated already in reference to the sorting of single leaves.

The hands are placed in well arranged heaps according to their species. The leaves are placed together in layers, flat, and in two rows, so that their points somewhat overlap towards the interior, whilst the strong ribs are turned outwards. The heaps of the planter should not be arranged higher than two or three feet so that the leaves in it may not become heated, especially they should, as a rule, be consigned to the buyer at a time when regular fermentation cannot have been completed.

But if the fermentation already commenced has to be interrupted on account of the delivery of the product, it is very prejudicial to its quality.

Superior leaves should have a layer beneath them of sandgut or refuse leaves, and a covering of the same on the top, in order to protect them from the injurious influences of the weather.

If the leaves have been stored in regularly arranged frames for the final hanging, they preserve there the requisite moisture and elasticity till the period of sorting and making into hands.

If this operation is performed with tobacco leaves dried and preserved in the usual manner, great difficulties have usually to be contended with, if at the time when these manipulations are undertaken it does not chance to be damp weather. If dry hard leaves are brought out of the cold into a warm room, the moisture strikes them, they absorb it, and the substance of the leaf becomes somewhat more flexible. But this simple means is seldom sufficient, so that if the occurrence of damp weather cannot be waited for, it becomes necessary to have recourse to artificial means by water or steam, which destroys the colour and induces decay. The least injurious means of preparing hard brittle leaves for bundling is to hang them up for a short time, perhaps over night, in a damp but not mouldy cellar till they attain the requisite flexibility.

If excessively moist leaves are subjected to bundling they stick together, especially at the points, soon pass into violent fermentation when stacked, become soft, of dark colour; and decay will soon appear if such bunches, immediately on delivery, are not at once submitted to drying. Bunches too wet at first and dried afterwards, always remain of inferior quality, and as the purchaser, on taking possession perceives this, it is evident that those planters who wet their tobacco leaves in order to increase their weight wilfully cause a considerable loss to themselves as well as to the purchaser.

In order to convey the hands into the delivery store without injury of form, there is, in Hungary, where the raising of tobacco is carried on on a large scale, a very simple mode of packing. For this purpose hard timbers are used, about two inches thick and five feet long, the first two pieces being placed on the ground, parallel and at a distance from each other some inches less than the breadth of the pile from which the leaves are to be removed.

These pieces of wood are then covered with a strip of mat called nassura, made of reeds, straw of Turkish millet, or merely laid with stalks of maize the breadth of the whole pile, and on this the hands are regularly built as before, in the schlicht. When the height is attained of from two feet to two and a half a similar cover of a nassura or one of maize stalks is laid on the top, in order to protect the surface of the leaves on all sides from wind or rain, and that they may not be injured by the timber.

The rib sides as a rule are not covered. On the top, in the centre of the pile, a post is laid lengthways, on which either several persons place themselves or by means of a simple lever power a pressure is exerted in order to compress the leaves to about the height of a third part of the pile. Too great a pressure is injurious to the unprepared leaves. They must therefore only be so much pressed as is necessary in order that the bale may not be afterward loose. After sufficient pressure has been exerted it must not be relaxed till the bale is ready. For this purpose there are placed on the top, along with the pole already there, two pieces of packing timber in such a manner that the edges somewhat project beyond the pile, and these timbers are fastened down as tightly as possible with strong knots, the slipping of which is guarded against by the ends of the wood being notched. If leaves for covering cigars are to be packed, a layer of refuse leaves is placed at the top and bottom, because the outermost layers sustain some injury through the strong pressure of the packing timbers, and become crumpled and stained with black marks.

The packing should be undertaken immediately before the delivery, because the leaves pressed into bales by lying long pass into fermentation.

TABLE showing the Amount and Composition of Ash existing in the Leaves and Stems of the Tobacco Plant.

Per-centage of ash of the dry substance.	I.	II.	III.	IV.	V.	VI.	VII.	VIII.	IX.		X.	
Leaves	18·9	19·8	23·0	21·1	23·3	23·3	22·8	27·3	—	14·5	—	18·9
Stems	22·9	32·5	19·8	—	—	—	—	—	9·2	—	10·7	—
Potash	29·1	18·2	8·2	19·5	9·7	9·3	10·3	11·2	47·4	18·1	47·8	25·9
Soda	2·2	—	—	0·3	—	—	—	—	6·0	—	6·0	5·8
Lime	27·7	27·8	42·8	44·5	49·3	49·4	39·5	47·0	—	12·2	—	20·4
Magnesia	7·2	15·7	13·9	11·1	14·6	15·6	15·0	12·8	—	9·8	0·5	2·0
Chloride of sodium	0·9	11·4	2·2	3·5	4·0	3·2	6·4	2·6	5·8	3·2	5·8	3·5
Chloride of potassium	—	3·9	8·5	—	4·4	3·3	3·0	3·0	—	5·6	—	—
Phosphate of iron	8·8	6·8	6·1	4·3	5·2	6·7	7·5	6·3	3·1	6·5	4·5	2·9
Phosphate of lime	—	—	—	6·0	—	—	—	—	25·2	16·1	26·0	18·1
Phosphate of magnesia	—	—	—	—	—	—	—	—	2·6	—	1·4	—
Sulphate of lime	6·4	10·1	8·0	5·6	6·7	6·1	9·4	5·1	—	16·7	—	5·9
Sulphate of soda	—	—	—	—	—	—	—	—	4·3	—	4·3	—
Silica	17·6	6·0	9·3	5·1	5·5	6·3	8·3	12·0	2·9	10·5	0·5	12·2
Sulphuric acid	—	—	—	—	—	—	—	—	—	—	—	—

Constituents of the ash in 100 parts of dry leaf.	Variety.							
	XI.	XII.	XIII.	XIV.	XV.	XVI.	XVII.	XVIII.
Potash	2·36	2·69	4·67	3·69	1·10	1·48	1·63	1·73
Lime	8·82	5·42	2·30	8·55	8·06	7·45	7·08	7.55
Magnesia	2·23	2·33	1·14	2·28	1·62	0·54	1·83	0·67
Oxide of iron	0·21	0·37	0·30	0·19	0·18	0·29	0·19	0·42
Silica	0·09	0·04	0·05	0·04	0·10	0·12	0·11	0·09
Chlorine	0·78	1·49	0·47	1·72	0·05	0·32	0·63	0·22
Sulphuric acid	0·52	0·41	0·52	0·82	0·83	0·86	0·82	0·98
Phosphoric acid	0·28	0·75	0·66	0·26	0·63	0·71	0·70	0·73
Sum	15·3	13·50	10·11	17·55	12·57	11·77	12·99	12·39

The analyses I.–VIII. inclusive are analyses of Hungarian tobaccos (air dry leaves and stems) from Funfkirchen, Debreczin, and the Banat, executed by Messrs. Will and Fresenius, quoted here from Mr. Mandis's treatise, page 31.

IX, a tobacco taken from the richest soil in Prince George's county, Maryland,
X, a tobacco taken from Hatfield, Connecticut river, Massachussets,
} analysed by Dr. Charles Jackson in Boston, quoted from the Patent Office Reports, 1858, p. 300.

XI, prepared tobacco for snuff from Funfkirchen (Hungary),
XII, prepared tobacco from Debrö (Hungary),
XIII, prepared tobacco from Virginia,
XIV, prepared tobacco from Szegedin (Hungary),
XV, XVI, XVII, XVIII, green tobacco leaves grown in Hungary from Mori, Palatinate, Virginia, and Debrö seed,
} analysed by Dr. Kodweiss, and quoted from Mr. Mandis's treatise, p. 32.

Remark. — The discrepancies between these different analyses are so great that it is hopeless to reconcile them.

The whole subject requires a new examination.

www.ingramcontent.com/pod-product-compliance
Lightning Source LLC
Chambersburg PA
CBHW062229220526
45471CB00009B/3404